Petroleum Engineering: Emerging Trends and Technologies

Petroleum Engineering: Emerging Trends and Technologies

Ripley Snyder

SYRAWOOD
PUBLISHING HOUSE

New York

Published by Syrawood Publishing House,
750 Third Avenue, 9th Floor,
New York, NY 10017, USA
www.syrawoodpublishinghouse.com

Petroleum Engineering: Emerging Trends and Technologies
Ripley Snyder

International Standard Book Number: 978-1-64740-122-1 (Hardback)

Cataloging-in-Publication Data

Petroleum engineering : emerging trends and technologies / Ripley Snyder.
    p. cm.
Includes bibliographical references and index.
ISBN 978-1-64740-122-1
1. Petroleum engineering. 2. Petroleum--Prospecting. 3. Mining engineering. I. Snyder, Ripley.
TN870 .P48 2022
665.5--dc23

Table of Contents

Preface   VII

Chapter 1   What is Petroleum Engineering?   1
    a.   Petroleum   1
    b.   Petroleum Engineering   19
    c.   Petroleum Geology   20
    d.   Hubbert Peak Theory   21
    e.   Octane Rating   22
    f.   Uses of Petroleum   23

Chapter 2   Branches of Petroleum Engineering   25
    a.   Reservoir Engineering   25
    b.   Petroleum Production Engineering   26
    c.   Subsurface Engineer   27
    d.   Drilling Engineering   28

Chapter 3   Methods and Techniques of Petroleum Exploration and Production   31
    a.   Petroleum Exploration   31
    b.   Petroleum Production   38
    c.   Hydrocarbon Exploration   65
    d.   Hydraulic Fracturing   70
    e.   Oil and Gas Exploration Methods   81
    f.   Spontaneous Potential Logging   82
    g.   Formation Evaluation Neutron Porosity   89
    h.   Well Logging   92
    i.   Fractional Distillation   101
    j.   Gamma Ray Logging   102
    k.   Deepwater Drilling   104
    l.   Directional Drilling   107
    m.   Pore Pressure Gradient   111

n. Pumpjack   113

o. Downhole Oil–Water Separation Technology   116

p. Drilling Mud   117

q. Mud Logging   118

Chapter 4 Petroleum Industry and Related Processes   129

a. Working of Oil and Gas Industry   130

b. Oil Refinery   132

c. Instrumentation in Petrochemical Industries   144

d. Oil Shale Industry   150

e. Petroleum Reservoir   155

f. Petroleum Refining Processes   158

g. Oil Well   164

Chapter 5 Byproducts of Petroleum Production   167

a. Paraffin Wax   167

b. Petrochemicals   168

c. Liquefied Petroleum Gas   175

d. Petroleum Coke   183

e. Crude Oil   188

f. Diesel Fuel   190

g. Petroleum Ether   191

h. Petroleum Jelly   193

i. Gasoline   198

j. Kerosene   211

Chapter 6 Environmental Impacts of Petroleum Production   213

a. Environmental and Economic Impacts of Crude Oil and
    Natural Gas Production   213

b. Environmental Impact of Hydraulic Fracturing   222

c. Environmental Impact of the Oil Shale Industry   230

d. Environmental Impact of the Petroleum Industry   233

Permissions

Index

Preface

This book has been written, keeping in view that students want more practical information. Thus, my aim has been to make it as comprehensive as possible for the readers. I would like to extend my thanks to my family and co-workers for their knowledge, support and encouragement all along.

The field of engineering which deals with the activities related to the production of hydrocarbons is known as petroleum engineering. Such hydrocarbons include crude oil or natural gas. Its main goal is to maximize the economic recovery of hydrocarbons from subsurface reservoirs. Petroleum engineering estimates the recoverable volume of the hydrocarbon resource with the help of a thorough understanding of the physical behavior of the resources. Such resources are found at very high pressure within the porous rocks. Petroleum engineering assimilates the knowledge of various other fields such as geophysics, petroleum geology, drilling, economics, formation evaluation, reservoir engineering, artificial lift systems, well engineering and petroleum production engineering. This textbook aims to shed light on some of the unexplored aspects of petroleum engineering. Most of the topics introduced in this book cover new techniques and the applications of this field. It is appropriate for those seeking detailed information in this area.

A brief description of the chapters is provided below for further understanding:

Chapter – What is Petroleum Engineering?

The field of engineering which is concerned with the activities related to the pro-duction of hydrocarbons, such as natural gas or crude oil, is referred to as petro-leum engineering. This is an introductory chapter which will introduce briefly all the significant aspects of petroleum engineering like petroleum geology, Hubbert peak theory and octane rating.

Chapter – Branches of Petroleum Engineering

Petroleum engineering is a vast field that can be divided into various branches. Some of them are reservoir engineering, petroleum production engineering, subsurface engineering and drilling engineering. The topics elaborated in this chapter will help in gaining a better perspective about these branches of petroleum engineering.

Chapter – Methods and Techniques of Petroleum Exploration and Production

Petroleum exploration deals with the exploration of deposits of petroleum. Various methods and techniques are used within this field such as hydraulic fracturing, spontaneous potential logging, formation evaluation neutron porosity, fractional distillation, etc. All these diverse techniques and methods related to the exploration and production of petroleum have been carefully analyzed in this chapter.

Chapter – Petroleum Industry and Related Processes

Petroleum industry deals with the processes of exploration, extraction, refining, transporting and marketing of petroleum products. It is related to various industries such as gas industry, oil shale industry and oil refinery. This chapter discusses in detail the petroleum industry and related processes like the petroleum refining process.

Chapter – Byproducts of Petroleum Production

Some of the most common byproducts of petroleum engineering are paraffin wax, petrochemicals, liquefied petroleum gas, petroleum coke, crude oil, diesel fuel, petroleum ether, petroleum jelly, gasoline and kerosene. This chapter has been carefully written to provide an easy understanding of these byproducts of petroleum production.

Chapter – Environmental Impacts of Petroleum Production

Due to the toxicity of petroleum it causes various negative environmental impacts such as air pollution, acid rain and illnesses in humans. The topics elaborated in this chapter will help in gaining a better perspective about the environmental impacts of different processes related to petroleum production such as crude oil and natural gas production, hydraulic fracturing, and the oil shale industry.

Ripley Snyder

1
What is Petroleum Engineering?

The field of engineering which is concerned with the activities related to the production of hydrocarbons, such as natural gas or crude oil, is referred to as petroleum engineering. This is an introductory chapter which will introduce briefly all the significant aspects of petroleum engineering like petroleum geology, Hubbert peak theory and octane rating.

Petroleum

Petroleum is a complex mixture of hydrocarbons that occur in Earth in liquid, gaseous, or solid form. The term is often restricted to the liquid form, commonly called crude oil, but, as a technical term, petroleum also includes natural gas and the viscous or solid form known as bitumen, which is found in tar sands. The liquid and gaseous phases of petroleum constitutethe most important of the primary fossil fuels.

Liquid and gaseous hydrocarbons are so intimately associated in nature that it has become customary to shorten the expression "petroleum and natural gas" to "petroleum" when referring to both.

The burning of all fossil fuels (coal and biomass included) releases large quantities of carbon dioxide (CO_2) into the atmosphere. The CO_2 molecules do not allow much of the long-wave solar radiation absorbed by Earth's surface to reradiate from the surface and escape into space. The CO_2 absorbs upward-propagating infrared radiation and re-emits a portion of it downward, causing the lower atmosphere to remain warmer than it would otherwise be. This phenomenon has the effect of enhancing Earth's natural greenhouse effect, producing what scientists refer to as anthropogenic (human-generated) global warming. There is substantial evidence that higher concentrations of CO_2 and other greenhouse gases have contributed greatly to the increase of Earth's near-surface mean temperature since 1950.

Exploitation of Surface Seeps

Small surface occurrences of petroleum in the form of natural gas and oil seeps have been known from early times. The ancient Sumerians, Assyrians, and Babylonians used crude oil, bitumen, and asphalt ("pitch") collected from large seeps at Tuttul (modern-day Hit) on the Euphrates for many purposes more than 5,000 years ago. Liquid oil was first used as a medicine by the ancient Egyptians, presumably as a wound dressing, liniment, and laxative. The Assyrians used bitumen as a means of punishment by pouring it over the heads of lawbreakers.

Oil products were valued as weapons of war in the ancient world. The Persians used incendiary arrows wrapped in oil-soaked fibres at the siege of Athens in 480 BCE. Early in the Common Era the Arabs and Persians distilled crude oil to obtain flammable products for military purposes. Probably as a result of the Arab invasion of Spain, the industrial art of distillation into illuminants became available in western Europe by the 12th century.

Several centuries later, Spanish explorers discovered oil seeps in present-day Cuba, Mexico, Bolivia, and Peru. Oil seeps were plentiful in North America and were also noted by early explorers in what are now New York and Pennsylvania, where American Indians were reported to have used the oil for medicinal purposes.

Extraction from Underground Reservoirs

Until the beginning of the 19th century, illumination in the United States and in many other countries was little improved over that which was known during the times of the Mesopotamians, Greeks, and Romans. Greek and Roman lamps and light sources often relied on the oils produced by animals (such as fish and birds) and plants (such as olive, sesame, and nuts). Timber was also ignited to produce illumination. Since timber was scarce in Mesopotamia, "rock asphalt" (sandstone or limestone infused with bitumen or petroleum residue) was mined and combined with sand and fibres for use in supplementing building materials. The need for better illumination that accompanied the increasing development of urban centres made it necessary to search for new sources of oil, especially since whales, which had long provided fuel for lamps, were becoming harder and harder to find. By the mid-19th century kerosene, or coal oil, derived from coal was in common use in both North America and Europe.

The Industrial Revolution brought an ever-growing demand for a cheaper and more convenient source of lubricants as well as of illuminating oil. It also required better sources of energy. Energy had previously been provided by human and animal muscle and later by the combustion of such solid fuels as wood, peat, and coal. These were collected with considerable effort and laboriously transported to the site where the energy source was needed. Liquid petroleum, on the other hand, was a more easily transportable source of energy. Oil was a much more concentrated and flexible form of fuel than anything previously available.

The stage was set for the first well specifically drilled for oil, a project undertaken by American entrepreneur Edwin L. Drake in northwestern Pennsylvania. The completion of the well in August 1859 established the groundwork for the petroleum industry and ushered in the closely associated modern industrial age. Within a short time, inexpensive oil from underground reservoirs was being processed at already existing coal oil refineries, and by the end of the century oil fields had been discovered in 14 states from New York to California and from Wyoming to Texas. During the same period, oil fields were found in Europe and East Asia as well.

Significance of Petroleum in Modern Times

At the beginning of the 20th century, the Industrial Revolution had progressed to the extent that the use of refined oil for illuminants ceased to be of primary importance. The oil and gas industry became the major supplier of energy largely because of the advent of the internal-combustion engine, especially those in automobiles. Although oil constitutes a major petrochemical feedstock, its primary importance is as an energy source on which the world economy depends.

The significance of oil as a world energy source is difficult to overdramatize. The growth in energy production during the 20th century was unprecedented, and increasing oil production has been by far the major contributor to that growth. By the 21st century an immense and intricate value chain was moving approximately 100 million barrels of oil per day from producers to consumers. The production and consumption of oil is of vital importance to international relations and has frequently been a decisive factor in the determination of foreign policy. The position of a country in this system depends on its production capacity as related to its consumption. The possession of oil deposits is sometimes the determining factor between a rich and a poor country. For any country, the presence or absence of oil has major economic consequences.

On a timescale within the span of prospective human history, the utilization of oil as a major source of energy will be a transitory affair lasting only a few centuries. Nonetheless, it will have been an affair of profound importance to world industrialization.

Properties of Hydrocarbons

Hydrocarbon Content

Although oil consists basically of compounds of only two elements, carbon and hydrogen, these elements form a large variety of complex molecular structures. Regardless of physical or chemical variations, however, almost all crude oil ranges from 82 to 87 percent carbon by weight and 12 to 15 percent hydrogen. The more-viscous bitumens generally vary from 80 to 85 percent carbon and from 8 to 11 percent hydrogen.

Crude oil is an organic compound divided primarily into alkenes with single-bond hydrocarbons of the form C_nH_{2n+2} or aromatics having six-ring carbon-hydrogen bonds,

C_6H_6. Most crude oils are grouped into mixtures of various and seemingly endless proportions. No two crude oils from different sources are completely identical.

The alkane paraffinic series of hydrocarbons, also called the methane (CH_4) series, comprises the most common hydrocarbons in crude oil. The major constituents of gasoline are the paraffins that are liquid at normal temperatures but boil between 40 °C and 200 °C (100 °F and 400 °F). The residues obtained by refining lower-density paraffins are both plastic and solid paraffin waxes.

The naphthenic series has the general formula C_nH_{2n} and is a saturated closed-ring series. This series is an important part of all liquid refinery products, but it also forms most of the complex residues from the higher boiling-point ranges. For this reason, the series is generally heavier. The residue of the refining process is an asphalt, and the crude oils in which this series predominates are called asphalt-base crudes.

The aromatic series is an unsaturated closed-ring series. Its most common member, benzene (C_6H_6), is present in all crude oils, but the aromatics as a series generally constitute only a small percentage of most crudes.

Nonhydrocarbon Content

In addition to the practically infinite mixtures of hydrocarbon compounds that form crude oil, sulfur, nitrogen, and oxygen are usually present in small but often important quantities. Sulfur is the third most abundant atomic constituent of crude oils. It is present in the medium and heavy fractions of crude oils. In the low and medium molecular ranges, sulfur is associated only with carbon and hydrogen, while in the heavier fractions it is frequently incorporated in the large polycyclic molecules that also contain nitrogen and oxygen. The total sulfur in crude oil varies from below 0.05 percent (by weight), as in some Venezuelan oils, to about 2 percent for average Middle Eastern crudes and up to 5 percent or more in heavy Mexican or Mississippi oils. Generally, the higher the specific gravity of the crude oil (which determines whether crude is heavy, medium, or light), the greater its sulfur content. The excess sulfur is removed from crude oil prior to refining, because sulfur oxides released into the atmosphere during the combustion of oil would constitute a major pollutant, and they also act as a significant corrosive agent in and on oil processing equipment.

The oxygen content of crude oil is usually less than 2 percent by weight and is present as part of the heavier hydrocarbon compounds in most cases. For this reason, the heavier oils contain the most oxygen. Nitrogen is present in almost all crude oils, usually in quantities of less than 0.1 percent by weight. Sodium chloride also occurs in most crudes and is usually removed like sulfur.

Many metallic elements are found in crude oils, including most of those that occur in seawater. This is probably due to the close association between seawater and the organic forms from which oil is generated. Among the most common metallic elements in oil are vanadium and nickel, which apparently occur in organic combinations as they do in living plants and animals.

Crude oil also may contain a small amount of decay-resistant organic remains, such as siliceous skeletal fragments, wood, spores, resins, coal, and various other remnants of former life.

Physical Properties

Crude oil consists of a closely related series of complex hydrocarbon compounds that range from gasoline to heavy solids. The various mixtures that constitute crude oil can be separated by distillation under increasing temperatures into such components as (from light to heavy) gasoline, kerosene, gas oil, lubricating oil, residual fuel oil, bitumen, and paraffin.

Crude oils vary greatly in their chemical composition. Because they consist of mixtures of thousands of hydrocarbon compounds, their physical properties—such as specific gravity, colour, and viscosity (resistance of a fluid to a change in shape)—also vary widely.

Specific Gravity

Crude oil is immiscible with and lighter than water; hence, it floats. Crude oils are generally classified as bitumens, heavy oils, and medium and light oils on the basis of specific gravity (i.e., the ratio of the weight of equal volumes of the oil and pure water at standard conditions, with pure water considered to equal 1) and relative mobility. Bitumen is an immobile degraded remnant of ancient petroleum; it is present in oil sands and does not flow into a well bore. Heavy crude oils have enough mobility that, given time, they can be obtained through a well bore in response to enhanced recovery methods—that is, techniques that involve heat, gas, or chemicals that lower the viscosity of petroleum or drive it toward the production well bore. The more-mobile medium and light oils are recoverable through production wells.

The widely used American Petroleum Institute (API) gravity scale is based on pure water, with an arbitrarily assigned API gravity of 10°. (API gravities are unitless and are often referred to in degrees; they are calculated by multiplying the inverse of the specific gravity of a liquid at 15.5 °C [60 °F] by 141.5). Liquids lighter than water, such as oil, have API gravities numerically greater than 10°. Crude oils below 22.3° API gravity are usually considered heavy, whereas the conventional crudes with API gravities between 22.3° and 31.1° are regarded as medium, and light oils have an API gravity above 31.1°. Optimum refinery crude oils considered the best are 40° to 45°, since anything lighter is composed of lower carbon numbers (the number of carbon atoms per molecule of material). Refinery crudes heavier than 35° API have higher carbon numbers and are more complicated to break down or process for optimal octane gasolines and diesel fuels. Early 21st-century production trends showed, however, a shift in emphasis toward heavier crudes as conventional oil reserves (that is, those not produced from source rock) declined and a greater volume of heavier oils was developed.

Boiling and Freezing Points

Because oil is always at a temperature above the boiling point of some of its compounds, the more volatile constituents constantly escape into the atmosphere unless confined. It is impossible to refer to a common boiling point for crude oil because of the widely differing boiling points of its numerous compounds, some of which may boil at temperatures too high to be measured.

By the same token, it is impossible to refer to a common freezing point for crude oil because the individual compounds solidify at different temperatures. However, the pour point—the temperature below which crude oil becomes plastic and will not flow—is important to recovery and transport and is always determined. Pour points range from 32 °C to below −57 °C (90 °F to below −70 °F).

Measurement Systems

In the United States, crude oil is measured in barrels of 42 gallons each; the weight per barrel of API 30° light oil is about 306 pounds. In many other countries, crude oil is measured in metric tons. For crude oil having the same gravity, a metric ton is equal to approximately 252 imperial gallons or about 7.2 U.S. barrels.

Formation Process

From Planktonic Remains to Kerogen: The Immature Stage

Although it is recognized that the original source of carbon and hydrogen was in the materials that made up primordial Earth, it is generally accepted that these two elements had to pass through an organic phase to be combined into the varied complex molecules recognized as hydrocarbons. The organic material that is the source of most hydrocarbons has probably been derived from single-celled planktonic (free-floating) plants, such as diatoms and blue-green algae, and single-celled planktonic animals, such as foraminifera, which live in aquatic environments of marine, brackish, or fresh water. Such simple organisms are known to have been abundant long before the Paleozoic Era, which began some 541 million years ago.

Rapid burial of the remains of the single-celled planktonic plants and animals within fine-grained sediments effectively preserved them. This provided the organic materials, the so-called protopetroleum, for later diagenesis (a series of processes involving biological, chemical, and physical changes) into true petroleum.

The first, or immature, stage of hydrocarbon formation is dominated by biological activity and chemical rearrangement, which convert organic matter to kerogen. This dark-coloured insoluble product of bacterially altered plant and animal detritus is the source of most hydrocarbons generated in the later stages. During the first stage, biogenic methane is the only hydrocarbon generated in commercial quantities. The production of biogenic methane gas is part of the process of decomposition of organic matter carried out by anaerobic microorganisms (those capable of living in the absence of free oxygen).

From Kerogen to Petroleum: The Mature Stage

Deeper burial by continuing sedimentation, increasing temperatures, and advancing geologic age result in the mature stage of hydrocarbon formation, during which the full range of petroleum compounds is produced from kerogen and other precursors by thermal degradation and cracking (in which heavy hydrocarbon molecules are broken up into lighter molecules). Depending on the amount and type of organic matter, hydrocarbon generation occurs during the mature stage at depths of about 760 to 4,880 metres (2,500 to 16,000 feet) at temperatures between 65 °C and 150 °C (150 °F and 300 °F). This special environment is called the "oil window." In areas of higher than normal geothermal gradient (increase in temperature with depth), the oil window exists at shallower depths in younger sediments but is narrower. Maximum hydrocarbon generation occurs from depths of 2,000 to 2,900 metres (6,600 to 9,500 feet). Below 2,900 metres, primarily wet gas, a type of gas containing liquid hydrocarbons known as natural gas liquids, is formed.

Approximately 90 percent of the organic material in sedimentary source rocks is dispersed kerogen. Its composition varies, consisting of a range of residual materials whose basic molecular structure takes the form of stacked sheets of aromatic hydrocarbon rings in which atoms of sulfur, oxygen, and nitrogen also occur. Attached to the ends of the rings are various hydrocarbon compounds, including normal paraffin chains. The mild heating of the kerogen in the oil window of a source rock over long periods of time results in the cracking of the kerogen molecules and the release of the attached paraffin chains. Further heating, perhaps assisted by the catalytic effect of clay minerals in the source rock matrix, may then produce soluble bitumen compounds, followed by the various saturated and unsaturated hydrocarbons, asphaltenes (precipitates formed from oily residues), and others of the thousands of hydrocarbon compounds that make up crude oil mixtures.

At the end of the mature stage, below about 4,800 metres (16,000 feet), depending on the geothermal gradient, kerogen becomes condensed in structure and chemically stable. In this environment, crude oil is no longer stable, and the main hydrocarbon product is dry thermal methane gas.

The Geologic Environment

Origin in Source Beds

Knowing the maximum temperature reached by a potential source rock during its geologic history helps in estimating the maturity of the organic material contained within it. This information may also indicate whether a region is gas-prone, oil-prone, both, or neither. The techniques employed to assess the maturity of potential source rocks in core samples include measuring the degree of darkening of fossil pollen grains and the colour changes in conodont fossils. In addition, geochemical evaluations can be made of mineralogical changes that were also induced by fluctuating paleotemperatures. In general, there appears to be a progressive evolution of crude oil characteristics from geologically younger, heavier, darker, more aromatic crudes to older, lighter, paler, more paraffinic types. There are, however, many exceptions to this rule, especially in regions with high geothermal gradients.

Accumulations of petroleum are usually found in relatively coarse-grained, permeable, and porous sedimentary reservoir rocks laid down, for example, from sand dunes or oxbow lakes; however, these rocks contain little, if any, insoluble organic matter. It is unlikely that the vast quantities of oil and natural gas now present in some reservoir rocks could have been generated from material of which no trace remains. Therefore, the site where commercial amounts of oil and natural gas originated apparently is not always identical to the location at which they are ultimately discovered.

Oil and natural gas is believed to have been generated in significant volumes only in fine-grained sedimentary rocks (usually clays, shales, or clastic carbonates) by geothermal action on kerogen, leaving an insoluble organic residue in the source rock. The release of oil from the solid particles of kerogen and its movement in the narrow pores and capillaries of the source rock is termed primary migration.

Accumulating sediments can provide energy to the migration system. Primary migration may be initiated during compaction as a result of the pressure of overlying sediments. Continued burial causes clay to become dehydrated by the removal of water molecules that were loosely combined with the clay minerals. With increasing temperature, the newly generated hydrocarbons may become sufficiently mobile to leave the source beds in solution, suspension, or emulsion with the water being expelled from the compacting molecular lattices of the clay minerals. The hydrocarbon molecules would compose only a very small part—a few hundred parts per million—of the migrating fluids.

Migration through Carrier Beds

The hydrocarbons expelled from a source bed next move through the wider pores of carrier beds (e.g., sandstones or carbonates) that are coarser-grained and more permeable. This movement is termed secondary migration and may be the result of rocks folding or raising from changes associated with plate tectonics. The distinction between primary and secondary migration is based on pore size and rock type. In some cases, oil may migrate through such permeable carrier beds until it is trapped by a nonporous barrier and forms an oil accumulation. Although the definition of "reservoir" implies that the oil and natural gas deposit is covered by more nonporous and nonpermeable rock, in certain situations the oil and natural gas may continue its migration until it becomes a seep on the surface, where it will be broken down chemically by oxidation and bacterial action.

Since nearly all pores in subsurface sedimentary formations are water-saturated, the migration of oil takes place in an aqueous environment. Secondary migration may result from active water movement or can occur independently, either by displacement or by diffusion. Because the specific gravity of the water in the sedimentary formation is considerably higher than that of oil and natural gas, both oil and natural gas will float to the surface of the water in the course of geologic time and accumulate in the highest portion of a trap. The collection under the trap is an accumulation of gas with oil and then formation water at the bottom. If salt is present in an area of weakness or instability near the trap, it can use the pressure difference between the rock and the fluids to intrude into the trap, forming a dome. The salt dome can be used as a subsurface storage vault for hazardous materials or natural gas.

Accumulation in Reservoir Beds

The porosity (volume of pore spaces) and permeability (capacity for transmitting fluids) of carrier and reservoir beds are important factors in the migration and accumulation of oil. Most conventional petroleum accumulations have been found in clastic reservoirs (sandstones and siltstones). Next in number are the carbonate reservoirs (limestones and dolomites). Accumulations of certain types of unconventional petroleum (that is, petroleum obtained through methods other than traditional wells) occur in shales and igneous and metamorphic rocks because of porosity resulting from fracturing. Porosities in reservoir rocks usually range from about 5 to 30 percent, but all available pore space is not occupied by petroleum. A certain amount of residual formation water cannot be displaced and is always present.

Reservoir rocks may be divided into two main types: (1) those in which the porosity and permeability is primary, or inherent, and (2) those in which they are secondary, or induced. Primary porosity and permeability are dependent on the size, shape, and grading and packing of the sediment grains and also on the manner of their initial consolidation. Secondary porosity and permeability result from postdepositional factors, such as solution, recrystallization, fracturing, weathering during temporary exposure at Earth's surface, and further cementation. These secondary factors may either enhance or diminish the initial porosity and permeability.

Traps

After secondary migration in carrier beds, oil and natural gas finally collect in a trap. The fundamental characteristic of a trap is an upward convex form of porous and permeable reservoir rock that is sealed above by a denser, relatively impermeable cap rock (e.g., shale or evaporites). The trap may be of any shape, the critical factor being that it is a closed inverted container. A rare exception is hydrodynamic trapping, in which high water saturation of low-permeability sediments reduces hydrocarbon permeability to near zero, resulting in a water block and an accumulation of petroleum down the structural dip of a sedimentary bed below the water in the sedimentary formation.

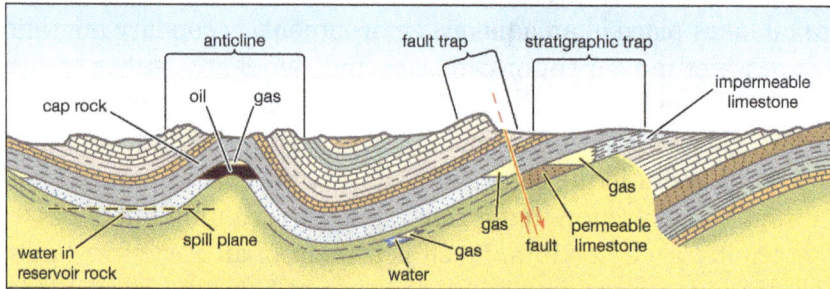

Principal types of petroleum traps.

Structural Traps

Traps can be formed in many ways. Those formed by tectonic events, such as folding or faulting of rock units, are called structural traps. The most common structural traps are anticlines, upfolds of strata that appear as inverted V-shaped regions on the horizontal planes of geologic maps. About 80 percent of the world's petroleum has been found in anticlinal traps. Most anticlines were produced by lateral pressure, but some have resulted from the draping and subsequent compaction of accumulating sediments over topographic highs. The closure of an anticline is the vertical distance between its highest point and the spill plane, the level at which the petroleum can escape if the trap is filled beyond capacity. Some traps are filled with petroleum to their spill plane, but others contain considerably smaller amounts than they can accommodate on the basis of their size.

Another kind of structural trap is the fault trap. Here, rock fracture results in a relative displacement of strata that form a barrier to petroleum migration. A barrier can occur when an impermeable bed is brought into contact with a carrier bed. Sometimes the faults themselves provide a seal against "updip" migration when they contain impervious clay gouge material between their walls. Faults and folds often combine to produce traps, each providing a part of the container for the enclosed petroleum. Faults can, however, allow the escape of petroleum from a former trap if they breach the cap rock seal.

Other structural traps are associated with salt domes. Such traps are formed by the upward movement of salt masses from deeply buried evaporite beds, and they occur along the folded or faulted flanks of the salt plug or on top of the plug in the overlying folded or draped sediments.

Stratigraphic Traps

A second major class of petroleum traps is the stratigraphic trap. It is related to sediment deposition or erosion and is bounded on one or more sides by zones of low permeability. Because tectonics ultimately control deposition and erosion, however, few stratigraphic traps are completely without structural influence. The geologic history of most sedimentary basins contains the prerequisites for the formation of stratigraphic traps. Typical examples are fossil carbonate reefs, marine sandstone bars, and deltaic distributary channel sandstones. When buried, each of these features provides a potential reservoir, which is often surrounded by finer-grained sediments that may act as source or cap rocks.

Sediments eroded from a landmass and deposited in an adjacent sea change from coarse-to fine-grained with increasing depth of water and distance from shore. Permeable sediments thus grade into impermeable sediments, forming a permeability barrier that eventually could trap migrating petroleum.

There are many other types of stratigraphic traps. Some are associated with the many transgressions (advances) and regressions (retreats) of the sea that have occurred over geologic time and the resulting deposits of differing porosities. Others are caused by processes that increase secondary porosity, such as the dolomitization of limestones or the weathering of strata once located at Earth's surface.

Resources and Reserves

Reservoirs formed by traps or seeps contain hydrocarbons that are further defined as either resources or reserves. Resources are the total amount of all possible hydrocarbons estimated from formations before wells are drilled. In contrast, reserves are subsets of resources; the sizes of reserves are determined by how economically or technologically feasible they are to extract petroleum from and use under current technological and economic conditions. Reserves are classified into various categories based on the amount that is likely to be extracted. Proven reserves have the highest certainty of successful extraction for commercial use (more than 90 percent), whereas successful extraction regarding probable and possible reserves for commercial use are estimated at 50 percent and between 10 and 50 percent respectively.

The broader category of resources includes both conventional and unconventional petroleum plays (or accumulations) as identified by analogs—that is, fields or reservoirs where there are few or no wells drilled but which are similar geologically to producing fields. For resources where some exploration or discovery activity has taken place, estimates of the size and number of undiscovered hydrocarbon accumulations are determined by technical experts and geoscientists as well as from measurements derived from geologic framework modeling and visualizations.

Unconventional Oil

Within the vast unconventional resources category, there are several different types of hydrocarbons, including very heavy oils, oil sands, oil shales, and tight oils. By the early

21st century, technological advances had created opportunities to convert what were once undeveloped resource plays into economic reserves.

Very heavy crudes have become economical. Those having less than 15° API can be extracted by working with natural reservoir temperatures and pressures, provided that the temperatures and pressures are high enough. Such conditions occur in Venezuela's Orinoco basin, for example. On the other hand, other very heavy crudes, such as certain Canadian crude oils, require the injection of steam from horizontal wells that also allow for gravity drainage and recovery.

Tar sands differ from very heavy crude oil in that bitumen adheres to sand particles with water. In order to convert this resource into a reserve, surface mining or subsurface steam injection into the reservoir must take place first. Later the extracted material is processed at an extraction plant capable of separating the oil from the sand, fines (very small particles), and water slurry.

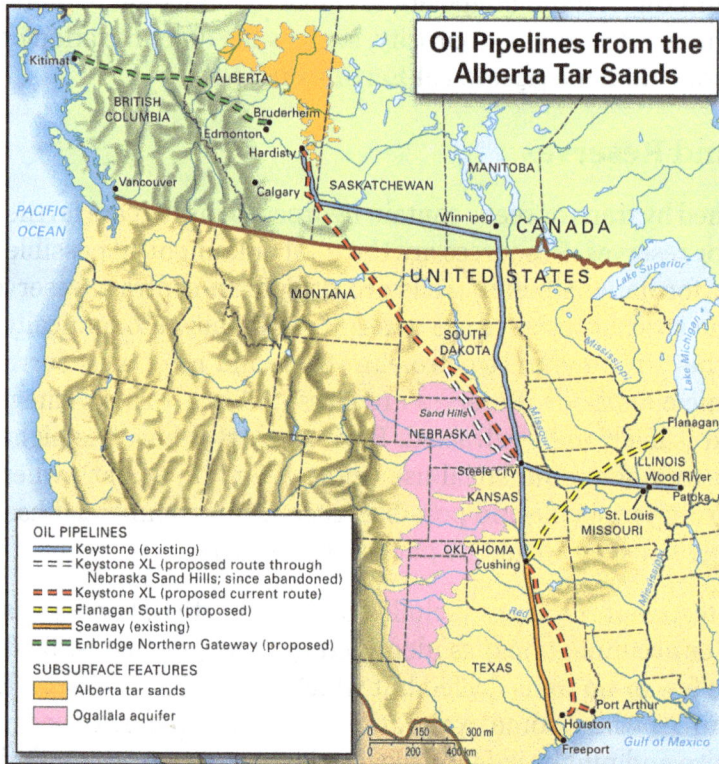

Alberta tar sands: The location of the Alberta tar sands region and its associated oil pipelines.

Oil shales make up an often misunderstood category of unconventional oils in that they are often confused with coal. Oil shale is an inorganic, nonporous rock containing some organic kerogen. While oil shales are similar to the source rock producing petroleum, they are different in that they contain up to 70 percent kerogen. In contrast, source rock tight oils contain only about 1 percent kerogen. Another key difference between oil shales and the tight oil produced from source rock is that oil shale is not exposed to sufficiently high temperatures to convert the kerogen to oil. In this sense, oil shales are hybrids of source rock oil and coal. Some oil shales can be burned as a solid. However, they are sooty and possess an extremely high volatile matter content when burned. Thus, oil shales are not used as solid fuels, but, after they are strip-mined and distilled, they are used as liquid fuels. Compared with other unconventional oils, oil shale cannot be extracted practically through hydraulic fracturing or thermal methods at present.

Shale oil is a kerogen-rich oil produced from oil shale rock. Shale oil, which is distinguished physically from heavy oil and tar sands, is an emerging petroleum source, and its potential was highlighted by the impressive production from the Bakken fields of North Dakota by the 2010s, which greatly boosted the state's petroleum output. (By 2015 North Dakota's daily petroleum production was approximately 1.2 million barrels, roughly 80 percent the amount produced per day by the country of Qatar, which is a member of Organization of the Petroleum Exporting Countries [OPEC]).

Tight oil is often light-gravity oil which is trapped in formations characterized by very low porosity and permeability. Tight oil production requires technologically complex drilling and completion methods, such as hydraulic fracturing (fracking) and other processes. (Completion is the practice of preparing the well and the equipment to extract petroleum). The construction of horizontal wells with multi-fracturing completions is one of the most effective methods for recovering tight oil.

Formations containing light tight oil are dominated by siltstone containing quartz and other minerals such as dolomite and calcite. Mudstone may also be present. Since most formations look like shale oil on data logs (geologic reports), they are often referenced as shale. Higher-productivity tight oil appears to be linked to greater total organic carbon (TOC; the TOC fraction is the relative weight of organic carbon to kerogen in the sample) and greater shale thickness. Taken together, these factors may combine to create greater pore-pressure-related fracturing and more efficient extraction. For the most productive zones in the Bakken, TOC is estimated at greater than 40 percent, and thus it is considered to be a valuable source of hydrocarbons.

Other known commercial tight oil plays are located in Canada and Argentina. For example, Argentina's Vaca Muerta formation was expected to produce 350,000 barrels per well when fully exploited, but by the early 21st century only a few dozen wells had been drilled, which resulted in production of only a few hundred barrels per day. In addition, Russia's Bazhenov formation in west Siberia has 365 billion barrels of recoverable reserves, which is potentially greater than either Venezuela's or Saudi Arabia's proved conventional reserves.

Considering the commercial status of all unconventional petroleum resource plays, the most mature reside within the conterminous United States, where unconventional petroleum in the liquid, solid, and gaseous phases is efficiently extracted. For tight oil, further technological breakthroughs are expected to unlock the resource potential in a manner similar to how unconventional gas has been developed in the U.S.

Unconventional Natural Gas

Perhaps the most-promising advances for petroleum focus on unconventional natural gas. (Natural gas is a hydrocarbon typically found dissolved in oil or present as a cap for the oil in a petroleum deposit). Six unconventional gas types—tight gas, deep gas, shale gas, coalbed methane, geopressurized zones, and Arctic and subsea hydrates—form the worldwide unconventional resource base. The scale of difference between conventional and unconventional reserves recoveries are commonly 30 percent to 1 percent, using tight gas as an example. In addition, the volume of the resource base is orders of magnitude higher; for example, 40 percent of all technically recoverable natural gas resources is attributable to shale gas. This total does not include tight gas, coalbed methane, or gas hydrates, nor does it include those shale gas resources that are believed to exist in unproven reserves in Russia and the Middle East.

World Distribution of Oil

Petroleum is not distributed evenly around the world. Slightly less than half of the world's proven reserves are located in the Middle East (including Iran but not North Africa). Following the Middle East are Canada and the United States, Latin America, Africa, and the region made up of Russia, Kazakhstan, and other countries that were once part of the Soviet Union.

The amount of oil and natural gas a given region produces is not always proportionate to the size of its proven reserves. For example, the Middle East contains approximately 50 percent of the world's proven reserves but accounts for only about 30 percent of global oil production (though this figure is still higher than in any other region). The United States, by contrast, lays claim to less than 2 percent of the world's proven reserves but produces roughly 16 percent of the world's oil.

Location of Reserves

Oil Fields

Two overriding principles apply to world petroleum production. First, most petroleum is contained in a few large fields, but most fields are small. Second, as exploration progresses, the average size of the fields discovered decreases, as does the amount of petroleum found per unit of exploratory drilling. In any region, the large fields are usually discovered first.

Since the construction of the first oil well in 1859, some 50,000 oil fields have been discovered. More than 90 percent of these fields are insignificant in their impact on world oil production. The two largest classes of fields are the supergiants, fields with 1 billion or more barrels of ultimately recoverable oil, and giants, fields with 500 million to 5 billion barrels of ultimately recoverable oil. Fewer than 40 supergiant oil fields have been found worldwide, yet these fields originally contained about one-half of all the oil so far discovered. The Arabian-Iranian sedimentary basin in the Persian Gulf region contains two-thirds of these supergiant fields. The remaining supergiants are distributed among the United States, Russia, Mexico, Libya, Algeria, Venezuela, China, and Brazil.

Although the semantics of what it means to qualify as a giant field and the estimates of recoverable reserves in giant fields differ between experts, the nearly 3,000 giant fields discovered—a figure which also includes the supergiants—account for 80 percent of the world's known recoverable oil. There are, in addition, approximately 1,000 known large oil fields that initially contained between 50 million and 500 million barrels. These fields account for some 14 to 16 percent of the world's known oil. Less than 5 percent of the known fields originally contained roughly 95 percent of the world's known oil.

Sedimentary Basins

Giant and supergiant petroleum fields and significant petroleum-producing basins of sedimentary rock are closely associated. In some basins, huge amounts of petroleum apparently have been generated because perhaps only about 10 percent of the generated petroleum is trapped and preserved. The Arabian-Iranian sedimentary basin is predominant because it contains more than 20 supergiant fields. No other basin has more than one such field. In 20 of the 26 most significant oil-containing basins, the 10 largest fields originally contained more than 50 percent of the known recoverable oil. Known world oil reserves are concentrated in a relatively small number of giant and supergiant fields in a few sedimentary basins.

Worldwide, approximately 600 sedimentary basins are known to exist. About 160 of these have yielded oil, but only 26 are significant producers, and 7 of these account for more than 65 percent of the total known oil. Exploration has occurred in another 240 basins, but discoveries of commercial significance have not been made.

Geologic Study and Exploration

Current geologic understanding can usually distinguish between geologically favourable and unfavourable conditions for oil accumulation early in the exploration cycle. Thus, only a relatively few exploratory wells may be necessary to indicate whether a region is likely to contain significant amounts of oil. Modern petroleum exploration is an efficient process. If giant fields exist, it is likely that most of the oil in a region will be found by the first 50 to 250 exploratory wells. This number may be exceeded if there is a much greater than normal amount of major prospects or if exploration drilling patterns are dictated by either political or unusual technological considerations. Thus, while undiscovered commercial oil fields may exist in some of the 240 explored but seemingly barren basins, it is unlikely that they will be of major importance since the largest are normally found early in the exploration process.

The remaining 200 basins have had little or no exploration, but they have had sufficient geologic study to indicate their dimensions, amount and type of sediments, and general structural character. Most of the underexplored (or frontier) basins are located in difficult environments, such as in polar regions, beneath salt layers, or within submerged continental margins. The larger sedimentary basins—those containing more than 833,000 cubic km (200,000 cubic miles) of sediments—account for some 70 percent of known world petroleum. Future exploration will have to involve the smaller basins as well as the more expensive and difficult frontier basins.

Status of the World Oil Supply

On several occasions—most notably during the oil crises of 1973–74 and 1978–79 and during the first half of 2008—the price of petroleum rose steeply. Because oil is such a crucial source of energy worldwide, such rapid rises in price spark recurrent debates about the accessibility of global supplies, the extent to which producers will be able to meet demand in the decades to come, and the potential for alternative sources of energy to mitigate concerns about energy supply and climate change issues related to the burning of fossil fuels.

How much oil does Earth have? The short answer to this question is that nobody knows. In its 1995 assessment of total world oil supplies, the U.S. Geological Survey (USGS) estimated that about 3 trillion barrels of recoverable oil originally existed on Earth and that about 710 billion barrels of that amount had been consumed by 1995. The survey acknowledged, however, that the total recoverable amount of oil could be higher or lower—3 trillion barrels was not a guess but an average of estimates based on different probabilities. This caveat notwithstanding, the USGS estimate was hotly disputed. Some experts said that technological improvements would create a situation in which much more oil would be ultimately recoverable, whereas others said that much less oil would be recoverable and that more than one-half of the world's original oil supply had already been consumed.

There is ambiguity in all such predictions. When industry experts speak of total "global oil reserves," they refer specifically to the amount of oil that is thought to be recoverable, not the total amount remaining on Earth. What is counted as "recoverable," however, varies from estimate to estimate. Analysts make distinctions between "proven reserves"—those that can be demonstrated as recoverable with reasonable certainty, given existing economic and technological conditions—and reserves that may be recoverable but are more speculative. The Oil & Gas Journal, a prominent weekly magazine for the petroleum industry, estimated in late 2007 that the world's proven reserves amounted to roughly 1.3 trillion barrels. To put this number in context, the world's population consumed about 30 billion barrels of oil in 2007. At this rate of consumption, disregarding any new reserves that might be found, the world's proven reserves would be depleted in about 43 years. However, because of advancements in exploration and unconventional oil extraction, estimates of the world's proven oil reserves had risen to approximately 1.7 trillion barrels by 2015.

By any estimation, it is clear that Earth has a finite amount of oil and that global demand is expected to increase. In 2007 the National Petroleum Council, an advisory committee to the U.S. Secretary of Energy, projected that world demand for oil would rise from 86 million barrels per day to as much as 138 million barrels per day in 2030. Yet experts remain divided on whether the world will be able to supply so much oil. Some argue that the world has reached "peak oil"—its peak rate of oil production. The controversial theory behind this argument draws on studies that show how production from individual oil fields and from oil-producing regions has tended to increase to a point in time and then decrease thereafter. "Peak-oil theory" suggests that once global peak oil has been reached, the rate of oil production in the world will progressively decline, with severe economic consequences to oil-importing countries.

A more widely accepted view is that through the early 21st century at least, production capacity will be limited not by the amount of oil in the ground but by other factors, such as geopolitics or economics. One concern is that growing dominance by nationalized oil companies, as opposed to independent oil firms, can lead to a situation in which countries with access to oil reserves will limit production for political or economic gain. A separate concern is that nonconventional sources of oil—such as oil sand reserves, oil shale deposits, or reserves that are found under very deep water—will be significantly more expensive to produce than conventional crude oil unless new technologies are developed that reduce production costs.

Petroleum System

Petroleum system starts with the deposition to storage from where the production is obtained. Petroleum system journey starts with the deposition of organic matter.

Deposition of Organic Matter

The deposition of organic matter starts when organism starts to die and deposits deep down the ocean floor and the above deposition of clay (finer grains). The clay particles are about 1/256 mm size and is called shale. Organic matter deposited on the ocean floor cannot be oxidized due to the depth factor so they can produce hydrocarbon. Hydrocarbon generation needs the cooking of organic matter at high temperature and pressure and it is obtained when it goes into overburden of deposition by clay particles and greater depths.

Source Rock

In the petroleum system the source rock are the shale (clay that goes under high pressure and temperature which cooks the organic matter). Sometimes limestone can also be the source rock with 1% of organic matter contains. So theses rocks undergo cooking where the temperature and pressure determines what type of fuel will be generated.

Despite of temperature and pressure another factor in producing hydrocarbon is the time span required to generate fuel. The time is a critical factor as if the organic matter is cooked for less time it will not generate hydrocarbon and when it is greater than the oil produce will be converted into gas.

Reservoir Rock

Reservoir as indicated by the name reserves of hydrocarbon. the hydrocarbon cannot be obtained from the source rock because of the higher pores but are lesser to none interconnection. For the extraction the pores should be interconnected so that it can travel when are extracted. But if there is no reservoir and obtaining fuel from source rock it must be fractured for permeability generation. Reservoir rock are mostly sandstone which have higher porosity and permeability but in some cases limestone also serves as reservoir rock. Limestone all by self is not a good reservoir due to fine particles present which give less permeability but as limestone is calcium carbonate so it can be dissolved in water which are the Karst topography. Only then can limestone have permeability required for hydrocarbon to be obtained.

Migration

Primary Migration

Migration itself is cleared so the primary migration occurs when hydrocarbon moves from source rock to the reservoir rocks. Primary migration occurs when the source rock is fractured due to tectonic forces (plate movements) or by the overburden squeezing the source rock. As HC (hydrocarbon) have low density they moves upward.

Secondary Migration

Secondary migration is the HC movement within the reservoir rocks. The HC will moves upward in the reservoir rocks.

Seal Rock

Trap or seal rocks are those that are present above reservoir rocks as HC movement will always be upward. Seal rock are those that have low to none permeability so that HC cannot escape but are trapped within the reservoir rocks. Shale can be seal rock also as they have porosity but do not have permeability factor so HC will be trapped. Types of traps include stratigraphic and structural. Stratigraphic traps example is shale as a seal rock and structural traps are fold or faults.

Time Period

The last thing in the petroleum system is time as have said it above already, time is required for HC generation which is always critical. No more time and no less time while cooking of the organic matter or it will not produce the fuel.

Petroleum Engineering

Petroleum engineering is the field of engineering that deals with the exploration, extraction, and production of oil. It also increasingly deals with the production of natural gas.

A petroleum engineer (also known as a gas engineer) determines the most efficient way to drill for and extract oil and natural gas at a particular well. They oversee drilling operations and resolve any operating problems. They also decide how to stimulate an underperforming well. They develop new drilling tools and techniques, and find new ways to extract remaining oil and gas from older wells.

Petroleum engineers evaluate oil and gas reservoirs to determine their profitability. They examine the geology of future drilling sites to plan the safest and most efficient method of drilling and recovering oil. They manage the installation, maintenance, and operation of equipment. They also manage the completion of wells. During production they monitor yield and develop modifications and stimulation programs to enhance it. They're also responsible for solving operational problems that may arise.

Since current extraction techniques only recover part of the available oil or gas in a reservoir, petroleum engineers also develop new drilling and extraction methods.

Petroleum engineers usually specialize in a particular aspect of drilling operations. For example, reservoir engineers determine the best way to recover the most oil or gas from a particular deposit, and estimate how much that will be. Drilling engineers figure out the best way to drill a particular well so that it's economically efficient and safe for people and the environment. Completions engineers decide the best way to finish building a well so that the oil or gas flows upwards from the ground. Production engineers monitor production and figure out how to coax more out of an under-producing well.

Petroleum engineers are vital to today's economies. They make the drilling process safer for people, communities, wildlife, and the environment. They also make it more efficient, and prices more affordable for customers. They ensure compliance with best practices, industry standards, and environmental and safety regulations, and contribute to energy independence.

Petroleum Geology

Petroleum geology is the application of geology (the study of rocks) to the exploration for and production of oil and gas. Geology itself is firmly based on chemistry, physics, and biology, involving the application of essentially abstract concepts to observed data. In the past, these data were basically observational and subjective, but they are now increasingly physical and chemical, and therefore more objective. Geology, in general, and petroleum geology, in particular, still rely on value judgments based on experience and an assessment of validity among the data presented.

Petroleum geology is essential for exploration and production and much more appropriate than random drilling. Petroleum geology and economic evaluation are critical pieces of prospect evaluation. Key geologic parameters in prospect appraisal are presence of:

- Source rock;

- Reservoir;

- Trap;

- Cap rock;

- Adequate and non-destructive thermal history.

The probability of each condition being fulfilled must be addressed. Four economic aspects are:

- Potential profitability of venture;

- Available risk investment;

- Total risk investment;

- Aversion to risk.

Computer simulation techniques may be used to aid the decision of whether or not to embark on an exploration venture.

Resources may be defined by two criteria: Economic feasibility of extraction and geologic knowledge. Reserves are resources that can be economically extracted from a resource base. Reserve categories can be subdivided into proved, probable, and possible.

Many estimates of global petroleum resources and remaining resources have been attempted and estimates differ widely. Most resources and reserves are contained in giant fields. The world consumes about 90 million barrels per day. Most estimates of world oil reserves indicate that this is about a 50-year supply of oil. Reserve numbers are always changing, however. In addition, the unconventional oil numbers have not been rigorously assessed.

Petroleum geology and its application to the reservoir process is critical for petroleum engineers. The geologist develops a viable model of the subsurface based on sparse observations in well logs, cores, and outcrops. The model needs to be consistent with geologic principles and based upon the model, exploration and development programs are planned. Petroleum systems occur in reservoirs within sedimentary basins in those areas of the world where subsidence of the earth's crust has allowed the accumulation of thick sequences of sedimentary rocks. Different types of petroleum traps such as anticlinal, fault, salt-related and stratigraphic traps and various types of reservoir rocks such as clastic (sandstone and shale), and carbonate rocks are enumerated. A seal rock keeps the oil entrapped in the reservoir and prevents it from migrating away. Understanding the geological and geomechanical nature of the seals is vital for successful exploration and reservoir development efforts.

Hubbert Peak Theory

Hubbert's peak theory is the idea that because oil production follows a bell-shaped curve, global crude oil production will eventually peak and then go into terminal decline. Although this model can be applied to many resources, it was developed as a model for oil production.

Breaking Down Hubbert's Peak Theory

Hubbert' peak theory is based on the work of Marion King Hubbert, a geologist working for Shell in the 1950s. It implies that maximum production from individual or global oil reserves will occur towards the middle of the reserve's life cycle — according to the Hubbert curve, which is used by exploration and production companies to estimate future production rates. After that, production decline accelerates due to resource depletion and diminishing returns. Accordingly, if new reserves are not brought online faster than extractable reserves are drawn down, the world will eventually reach peak oil — because there is a finite amount of conventional light, sweet crude in the earth's crust.

A Technological Revolution in Oil Production

But Hubbert's predictions that U.S. oil production would peak in the 1970s, and that the world would hit peak oil around the year 2000, were proven wrong, because a technological revolution in the oil business has increased recoverable reserves, as well as boosting recovery rates from new and old wells.

Thanks to high-tech digital oil exploration using 3D seismic imaging that enables scientists to see miles below the seabed floor, proven reserves around the world are growing all the time, as new oil fields are discovered. Offshore drilling in the 1950s could reach a depth of 5,000 feet. Today it is 25,000 feet.

The U.S. exceeded its former 1972 peak of 10.2 million barrels per day in January 2018, thanks to innovations like hydraulic fracturing, enhanced oil recovery and horizontal drilling. This has added trillions of cubic feet of gas and billions of barrels of oil to America's recoverable reserves, and turned it into a net exporter of petroleum products.

The oil industry no longer talks about running out of oil, thanks to companies like Schlumberger. For the foreseeable future, there are almost unlimited quantities of oil. Technically, recoverable oil is estimated to be around 2.6 trillion barrels and rising, because most of the world has yet to be explored using the latest technologies.

Nor are we anywhere close to peak energy. There are an estimated 1.1 trillion tonnes of proven coal reserves worldwide — enough to last around 150 years at current rates of production. There are 205.34 trillion cubic meters of proven natural gas reserves — enough to last at least 50 years. And there may be 3,000 billion tonnes of methane hydrates — which is enough natural gas to fuel the world for a thousand years, according to the U.S. Geological and Geophysical Service.

Octane Rating

The octane number is a value used to indicate the resistance of a motor fuel to knock. Octane number is also known as octane rating. Octane numbers are based on a scale on which isooctane is 100 (minimal knock) and heptane is 0 (bad knock). The higher the octane number, the more compression required for fuel ignition. Fuels with high octane numbers are used in high performance gasoline engines. Fuels with low octane number (or high cetane numbers) are used in diesel engines, where fuel is not compressed.

Octane Number Example

A gasoline with an octane number of 92 has the same knock as a mixture of 92% isooctane and 8% heptane.

Why the Octane Number Matters?

In a spark-ignition engine, using a fuel with too low an octane rating can lead to pre-ignition and engine knock, which can cause engine damage. Basically, compressing the air-fuel mixture may cause fuel to detonate before the flame front from the spark plug reaches it. The detonation produces higher pressure than the engine may be able to withstand.

Uses of Petroleum

After water, petroleum is another liquid that humans are most dependent upon. It is a source of fuel and is mostly found in many consumer products. To begin with, petroleum is a naturally occurring element which is in liquid state. It is either yellow or black in color. It is usually found in geological formations underneath the earth's crust. Talking about the composition, petroleum is a mix of thousands of molecules and organic compounds. However, hydrocarbons of various molecular weights form the majority of them.

Moreover, the most prolific hydrocarbons found in the chemistry of petroleum are alkanes. It also contains cycloalkanes, aromatic hydrocarbons more complex hydrocarbons such as asphaltenes.

Petroleum is mostly recovered by oil drilling while its constituents are separated using a process called fractional distillation. The term petroleum covers both unprocessed crude oil and other products that are made from refined crude oil.

Different uses of Petroleum

When we talk about petroleum and its uses, most of the people generally think of it as fuel either petrol or diesel. However, petroleum has been used in one form or another. It is an important substance across society, politics, technology including in economy. Besides, apart from fuel there are a lot of petroleum byproducts that show up in our modern life. Let's look at some of the uses of petroleum below:

- Agriculture.
- Detergents, Dyes, and Others.
- Plastics, Paints and More.
- Pharmaceuticals.
- Rubber.

Agriculture

When we talk about agriculture we are talking about fertilizers. Here, petroleum is used in the production of ammonia which serves a source of nitrogen. The Haber process is used in this case. Pesticides are also made from oil. All in all, petroleum based products are used extensively in agriculture as it helps in running farm machinery and fertilize plants.

Detergents, Dyes and Others

Distillates of petroleum that include toluene, benzene, xylene, amongst others are used to obtain raw materials that are further used in products like synthetic detergents, dyes, and fabrics. Benzene and toluene which gives polyurethanes is often used in oils or surfactants, and it is also used to varnish wood.

Plastics, Paints and More

Plastics are mostly made of petrochemicals. Petroleum-based plastic like nylon or Styrofoam and other are made from this element. Usually, the plastics come from olefins, which include ethylene and propylene. Petrochemicals are also used to produce oil based paints or paint additives. Petrochemical ethylene is found in photographic film.

Pharmaceuticals and Cosmetics

Petroleum by-products like mineral oil and petrolatum are used in many creams and other pharmaceuticals. Tar is also produced from petroleum. Cosmetics that contain oils, perfumes are petroleum derivatives.

Rubber

Petrochemicals are also used in manufacturing synthetic rubber which is further used to make rubber soles on shoes, car tire and others rubber products. Rubber is primarily a product of butadiene.

Popular Products made from Petroleum

Some of the products made from or contain petroleum are; wax, ink, vitamin capsule, denture adhesive, toilet seats, upholstery, CDs, putty, guitar strings, crayons, pillows, artificial turf, hair coloring, deodorant, lipstick, heart valves, anesthetics cortisone, aspirin.

References

- Petroleum, science: britannica.com, Retrieved 13 March, 2019

- Petroleum-system: geologylearn.blogspot.com, Retrieved 14 April, 2019

- Petroleum-engineer, career: nvironmentalscience.org, Retrieved 15 May, 2019

- Petroleum-geology, earth-and-planetary-sciences, topics: sciencedirect.com, Retrieved 16 June, 2019

- Hubbert-peak-theory: terms, investopedia.com, Retrieved 17 July, 2019

- Definition-of-octane-number-604586: thoughtco.com, Retrieved 18 August, 2019

- Uses-of-petroleum, chemistry: byjus.com, Retrieved 19 January, 2019

2
Branches of Petroleum Engineering

Petroleum engineering is a vast field that can be divided into various branches. Some of them are reservoir engineering, petroleum production engineering, subsurface engineering and drilling engineering. The topics elaborated in this chapter will help in gaining a better perspective about these branches of petroleum engineering.

Reservoir Engineering

Reservoir engineering involves more than applied reservoir mechanics. The objective of engineering is optimization. To obtain optimum profit from a field the engineer or the engineering team must identify and define all individual reservoirs and their physical properties, deduce each reservoir's performance, prevent drilling of unnecessary wells, initiate operating controls at the proper time, and consider all important economic factors, including income taxes. Early and accurate identification and definition of the reservoir system is essential to effective engineering. Conventional geologic techniques seldom provide sufficient data to identify and define each individual reservoir; the engineer must supplement the geologic study with engineering data and tests to provide the necessary information. Reservoir engineering is difficult. The most successful practitioner is usually the engineer who, through extensive efforts to understand the reservoir, manages to acquire a few more facts and thus needs fewer assumptions.

Reservoir engineering has advanced rapidly during the last decade. The industry is drilling wells on wider spacing, unitizing earlier, and recovering a greater percentage of the oil in place. Techniques are better, tools are better, and background knowledge of reservoir conditions has been greatly improved. In spite of these general advances, many reservoirs are being developed in an inefficient manner, vital engineering considerations often are neglected or ignored, and individual engineering efforts often are inferior to those of a decade ago. Reservoir engineers often disagree in their interpretation of a reservoir's performance. It is not uncommon for two engineers to take exactly opposite positions before a state commission. Such disagreements understandably confuse and bewilder management, lawyers, state commission members and laymen. Can they be blamed if they question the technical competence of a professional group whose members cannot agree among themselves? There is considerable difference between the reservoir engineering practiced by different companies. The differences between good engineering and ineffective engineering generally involve only minor variations in fundamental knowledge but involve major differences in emphasis of what is important. Some companies or groups emphasize calculation procedures and reservoir mechanics, but pay little attention to reservoir geology. Others emphasize geology and make extensive efforts to identify individual reservoirs and deduce their performance during the development period or during the early operating period. They use reservoir engineering equations and calculation procedures primarily as tools to provide additional insight of a reservoir's performance. Those utilizing the latter approach generally are the most successful. The differences in practice observed indicate that many individuals, including managers, field personnel, educators, scientists and reservoir engineers do not understand the full scope of reservoir engineering or bow the reservoir engineer can be used most effectively. A better understanding of the basic purpose of reservoir engineering and how it can be utilized most effectively should result in improved engineering.

The purpose of engineering the goal of engineering is optimization. The purpose of reservoir engineering is to provide the facts, information and knowledge necessary to control operations to obtain the maximum possible recovery from a reservoir at the least possible cost. Since a maximum recovery generally is not obtained by a minimum expenditure, the engineer must seek some optimum combination of recovery, cost, and other pertinent factors. How one defines "optimum" will depend upon the policies of the various operators. From an operator's point of view any procedure or course of action that results in an optimum profit to the company is effective engineering, and any that doesn't is not. There are two reasons why a company may not receive effective engineering. Its engineers may be poorly trained and fail to perform property. However, a company can employ competent engineers and receive good engineering work from them, but as a company, still do an ineffective job of engineering. For instance, an engineer might do an excellent job of water flooding a reservoir. However, if even greater profit could have been received by water flooding five years earlier, then obviously the reservoir was not effectively engineered by the operator.

Petroleum Production Engineering

Petroleum production engineering is a subset of petroleum engineering. Petroleum production engineers design and select subsurface equipment to produce oil and gas well fluids. They often are degreed as petroleum engineers, although they may come from other technical disciplines (e.g., mechanical engineering, chemical engineering, physicist) and subsequently be trained by an oil and gas company.

Petroleum production engineers' responsibilities include:

1. Evaluating inflow and outflow performance between the reservoir and the wellbore.

2. Designing completion systems, including tubing selection, perforating, sand control, matrix stimulation, and hydraulic fracturing.

3. Selecting artificial lift equipment, including sucker-rod lift (typically beam pumping), gas lift, electrical submersible pumps, subsurface hydraulic pumps, progressing-cavity pumps, and plunger lift.

4. Selecting (not design) equipment for surface facilities that separate and measure the produced fluids (oil, natural gas, water, and impurities), prepare the oil and gas for transportation to market, and handle disposal of any water and impurities.

Surface equipments are designed by Chemical engineers and Mechanical engineers according to data provided by the production engineers.

Subsurface Engineer

Subsurface engineers (also known as "completion engineers") are a subset within Petroleum Engineering and typically work closely with Drilling engineers. The job of a Subsurface Engineer is to effectively select equipment that will best suit the subsurface environment in order to best produce the hydrocarbon reserves. Once the hardware has been selected, a Subsurface Engineer will monitor and adjust the equipment to ensure the well and reservoir produces under ideal circumstances.

Subsurface engineers must design a successful well completion system by selecting equipment that is adequate for both downhole environments and applications. Considerations must be given to the various functions under which the completion equipment must operate and the effects any changes in temperatures or differential pressure will have on the equipment. The completion system must also be efficient and cost effective to achieve maximum production and financial goals. Another factor in the selection of specific completion equipment is the production rates of the well. The typical job duties of a Subsurface engineer include managing the interface between the reservoir and the well, including perforations, sand control, artificial lift, downhole flow control, and downhole monitoring equipment. Additional responsibilities of a Subsurface engineer include: performing a cost and risk analysis on the design, contacting vendors for the rental, purchase, and shipment of equipment, and working closely with fellow employees (geologists, reservoir engineers, drilling engineers, and production engineers).

The Society of Petroleum Engineers (SPE) has technical disciplines which allow SPE members to focus their attention on the technical activities that most interest them.

Drilling and Completions historically have been intertwined work within Petroleum Engineering. In 2016, SPE split the Drilling and Completions technical disciplines so SPE members would be able to focus more on Drilling or Completions. SPE continues to publish the SPE Drilling & Completions journal, it has been publishing the journal since 1993. SPE illustrates the technical activities of Drilling and Completions on its website and also hosts a page about SPE offerings related to Completions engineering. SPE also has many on demand webinars on Completions topics.

Design Components

The design components considered to perform a well completion may include:

- Cost and risk analysis.
- Determining the specifications for the wellbore clean-out.
- Use of specific Packer assemblies.
- Determining specific tool selection to operate equipment within the well.
- Assess possible equipment load specifications and incorporation of safety factors.
- Best use of flow control accessories (sliding sleeves and safety valves).
- Determining the appropriate perforating shots per foot and charges based on the target formations.
- Acidifying the formation to inhibit flow of hydrocarbons.
- Sand Control operations to increase production.
- Prevention of formation sand production with the use of wire screens.
- Review Well logs to determine equipment placement within the well.
- Determination of specific production pipe regarding well flow rates.
- Selection of equipment to maintain well stability.
- Oversee completion operations.

Drilling Engineering

Drilling engineering is the science behind the wells that produce oil and gas. Drilling engineering involves the planning, costing, developing and supervising of oil and gas well operations. Drilling engineering usually involves temporarily intense projects related to well design, testing and completion. The science of drilling engineering is divided into the four different activities below:

Completion Engineering

Completion engineering concerns the development of plans to improve the production from gas and oil wells. Completion engineers design, monitor and report on the installation of wells. They devise and discuss ways and methods to improve oil and gas well production. Completion engineers must plan delivery timetables, track the movement of products through warehouses and monitor local stock and equipment levels. They oversee and coordinate the arrival of shipments in order to streamline operations. Completion engineers may work closely with supply chain managers and even supervise warehouse staff. Completion engineering requires travel to offshore and remote locations.

Operational Engineering

Operational engineers are responsible for day-to-day well planning and installation operations. They may work specifically with corporate testing, safety, environmental and industry standards programs. Operational engineers ensure that data is accurately collected in order to present professional reports to clients and management. They usually liaison with the onshore operations supervisor to ensure they are fully aware of ongoing or upcoming issues.

They may attend pre- and post-test meetings with clients and sometimes direct on-site client tours. Operational engineers sometimes conduct company health and safety training while also reviewing and updating policies and procedures. Operational engineers work directly with production engineers.

Production Engineering

A production engineer designs and selects the tools and equipment that will cause the well to produce oil and gas after drilling. They usually have an academic background in mechanical and geosystems engineering. Production engineers coordinate the purchase, installation, maintenance and operation of the mining or oilfield equipment. They may even manage the interconnected operations between the well and the reservoir using things like sand control, artificial lifts and special hole controls. During this time, they inspect the well to ensure that oil or gas is safely and optimally flowing. They are expected to recommend modifications to maximize the efficiency of oil and gas recovery while maintaining economic viability.

Reservoir Engineering

Reservoir engineering involves the assessments of oil and gas deposits. Reservoir engineers are the professionals who estimate the potential size of the reservoir in order to determine how much oil and gas is available. Based on their calculations, they decide how to maximize the return on interest and operational efficiency. Because it's almost always impossible physically view subsurface fluids, reservoir engineers must work together with geologists, geo-hydrologists and geosystems engineers to accurately locate the oil and gas reserves through the advanced laws of physics and chemistry. They may conduct experiments involving the study of the behavioral effects of oil, water and natural gas in rocky subsurface settings.

Anyone who wants to become a drilling engineer will most likely need to pursue a degree in petroleum engineering, which will cover the principles of science, engineering and mathematics as they relate to oil and gas drilling, production and maintenance. These degrees may include courses in mechanics, geostatistics, well testing, hydro-geology and thermodynamics. These degrees will probably include classes on project and drilling operations management.

References

- "SPE Splits Drilling & Completion Disciplines as Operators Increase Focus on Completion Technologies - wellez". Wellez. 2016-06-07. Retrieved 2018-03-13.

- Journal-paper, SPE-920-PA: onepetro.org, Retrieved 20 February, 2019

- Clegg, Joe Dunn, ed. (2007). Petroleum-Engineering-Handbook-Volume-IV-Production-Operations-Engineering. Dallas, Texas: Society of Petroleum Engineers. P. 900. ISBN 978-1-55563-118-5

- What-is-drilling-engineering, faq: greatvaluecolleges.net, Retrieved 21 March, 2019

- Career, My Oil and Gas. "Completions Engineer Oil & Gas Careers | myoilandgascareer.com". Www.myoilandgascareer.com. Retrieved 2018-01-31

3
Methods and Techniques of Petroleum Exploration and Production

Petroleum exploration deals with the exploration of deposits of petroleum. Various methods and techniques are used within this field such as hydraulic fracturing, spontaneous potential logging, formation evaluation neutron porosity, fractional distillation, etc. All these diverse techniques and methods related to the exploration and production of petroleum have been carefully analyzed in this chapter.

Petroleum Exploration

The role of exploration is to provide the information required to exploit the best opportunities presented in the choice of areas, and to manage research operations on the acquired blocks.

An oil company may work for several years on a prospective area before an exploration well is spudded and during this period the geological history of the area is studied and the likelihood of hydrocarbons being present quantified.

	1	2	3	4	5	6
Initial Evalution	▬▬					
Geological Survey		▬▬				
Geophysical Survey			▬▬		▬	
Drilling				▬		▬▬
Decisions				↑ drill	↑ continue?	↑ continue?

Stages of a typical exploration program.

Exploration is responsible for handling the risk intrinsic in this activity, and this is generally achieved by selection of a range of options in probabilistic and economic terms.

Indeed, exploration is a risk activity and the management of exploration assets and associated operations is a major task for oil companies.

The risk cannot be eliminated entirely but can be controlled and reduced adopting appropriate workflow, conceptual and technological innovations.

When it's been decided to start up with an exploration project in a basin or in a larger area containing several basins, the quantity and quality of available data must be acquired and evaluated – geological data, type of reserves, production of existing fields (if any), etc.

Basin assessment/evaluation is the first step to undertake the study of the area under interest.

Technological development has provided oil companies with Basin Modeling – which is a numerical simulations that allows the temporal reconstruction of the history of a sedimentary basin and the associated evolution of the processes related to the formation of petroleum accumulations.

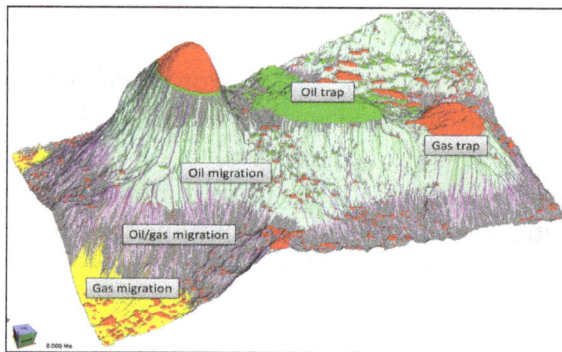
Basin modeling – Petroleum system.

On the basis of data and evidences collected from the preliminary studies, the company management, in the light of the possibilities and the probabilities of a discovery based on G&G data, aside from considerations of an economic nature, may decide to move to the following stage, which is the acquisition (through direct negotiations or by taking part in bids, etc.) of the legal right to perform prospecting in the selected area/block.

The owner of the mining right is normally the State, with which the oil company stipulates a contract establishing the contracting parties' rights.

Production Sharing Contracts and service contracts are frequently adopted nowadays.

The sequence of activities covered by an exploration permit is fairly uniform, and include:

- The creation of a database.
- The analysis of available data.
- The programming of mapping and geological and photo-geological surveys.
- Seismic surveys and interpretation of seismic data.
- The choice of well locations, drilling.

- The analysis of results and the decision as to whether or not to proceed with the application for a lease or to release the area after fulfilling obligations.

Goal of exploration is to identify and locate a prospect, to quantify the volume of hydrocarbon which might be contained in the potential reservoirs and to evaluate the risk inherent the project itself.

A prospect is a viable target evidenced by geological and geophysical indications that is recommended for drilling an exploration well.

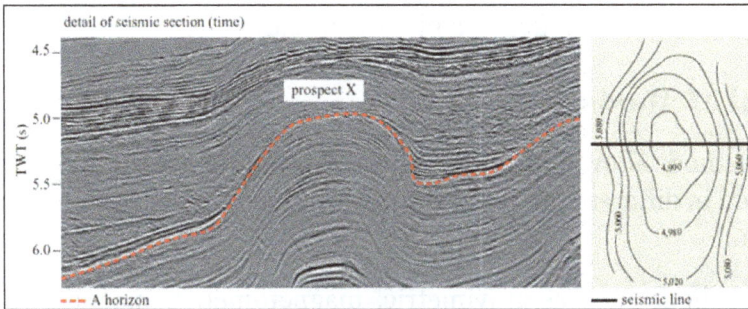

A detail of seismic section (time) and A horizon
depth map (meters) referred to prospect X.

The prospects identified must be technically practicable and meet the market conditions to guarantee a financial return on investments. The results obtained by drilling the exploratory wells indicate whether the initial geological hypotheses are correct or whether variations are found.

All this will allow the fine-tuning of the economic analysis of the project possibly turning hypothetical reserves into proved ones.

Where profitability does not meet the standards of the company, it leads to the termination of further investments.

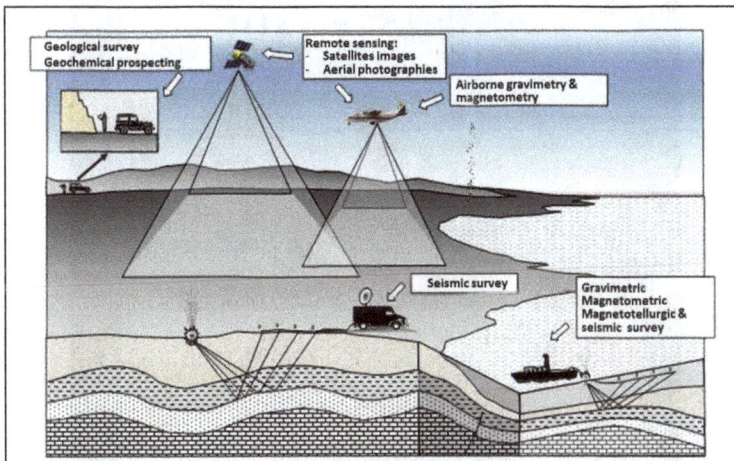

The main petroleum exploration techniques.

Geological Mapping and Prospecting

Geological mapping and prospecting are valuable techniques in an petroleum exploration. It is basically a technique which allows a graphical presentation of geological observations and interpretations.

Geological prospecting make use of geological disciplines such as petrography, stratigraphy, sedimentology, structural geology, geochemistry.

Such disciplines are used to achieve different targets but it must be stressed that their integration is fundamental to depict a picture of reality.

Geophysical Methods

Geophysical methods allow to study the physical properties of the subsurface rocks and they can be used in different phases of the exploration in order to collect different types of information.

Geophysical methods such as gravimetric, magnetometric, magnetotelluric, seismic are often combined to obtain more accurate and reliable results.

- Gravimetric prospecting:

  ◦ Gravimetric prospecting is a geophysical technique which is able to identify anomalies in the gravity acceleration generated by contrasts in density among bodies in the subsurface.

  ◦ Gravimetric prospecting is used to reconstruct of the main structural elements of sedimentary basins such as:

    ▪ Extension, thickness, salt domes, intrusive plutons and dislocations or fault lines.

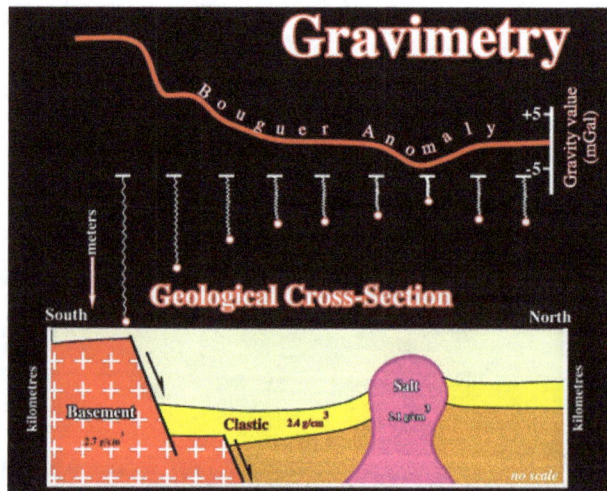

- Magnetometric Prospecting:

  ◦ This method involves measuring local anomalies in the Earth's magnetic fields.

  ◦ The method enable acquisition of data on structural characteristics and depth of the susceptive basement and therefore, indirectly, on the thickness of sedimentary overburden and identifies the presence, depth and extension of volcanic or plutonic masses within the sedimentary sequences.

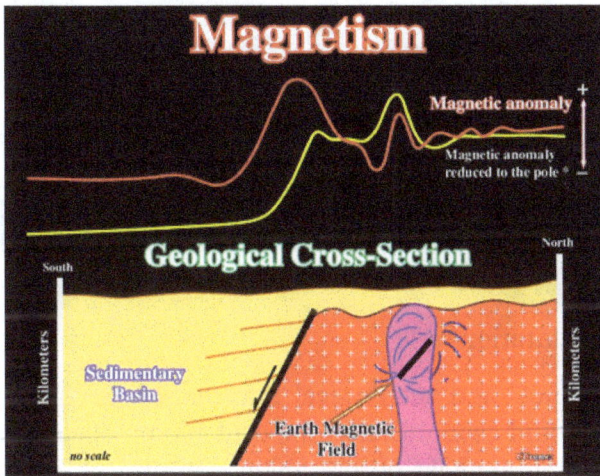

- Seismic Prospecting:

  ◦ Seismic prospecting has become the most valuable technique to reduce exploration risk of being unsuccessful in locating a prospect.

  ◦ The technique is based on determinations of the time interval that elapses between the initiation of a seismic wave at a selected shop point and the arrival of reflected or refracted impulses at one or more seismic detectors.

  ◦ The phase of seismic data acquisition is followed by the seismic data processing phase (aimed to the alteration of seismic data to suppress noise, enhance signal and migrate seismic events to the appropriate location in space) than by the interpretation of the generated subsurface image.

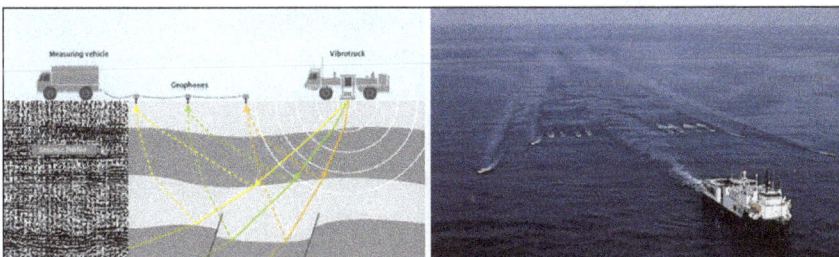

Onshore seismic survey.   Marine seismic survey.

Geophysicists interpret the processed seismic data and integrate other geoscientific information to make assessments of where oil and gas reservoirs may be accumulated.

Powered by advanced supercomputer power, rapid data loading, high-speed networking and high-resolution graphics, visualization centers provide the ability to display and manipulate complex volumes of 3D data resulting in better interpretation of more data in less time.

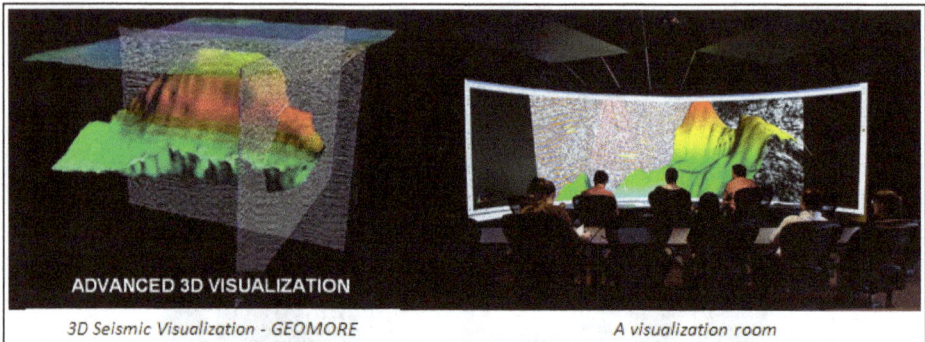

ADVANCED 3D VISUALIZATION

3D Seismic Visualization - GEOMORE   A visualization room

- Drilling the exploration well:

    ○ Once geological and geophysical information have defined and evaluated (technically and economically) the drillable prospect, it is possible to move to a fundamental phase of the exploration project – the drilling of the first exploratory well.

    ○ The drilling of the exploration well is aimed to confirm the presence of the petroleum accumulation.

Onshore and offshore exploration well.

- Well logging:

    ○ The well logging technique consists of lowering a 'logging tool' into the well to acquire geological data and to reveal reservoir fluids characteristics.

    ○ Well logging help geoscientists and engineers to understand:

        ▪ Presence of reservoir.

- ▪ Presence of hydrocarbons and characteristics.

  ◦ Reservoir properties, etc.

- Coring:

  ◦ Coring provides the only means of obtaining highquality samples for the direct measurement of rock and reservoir properties.

  ◦ Core samples are then used to perform Routine Core Analysis and Special Core Analysis to obtain detailed petrophysical data.

Coring and coring equipment at surface.

- Well Testing:

  ◦ A well test is the measurement under controlled conditions of all factors relating to the production of oil, gas, and water from a well.

  ◦ Well tests are conducted to acquire dynamic rate, pressure, temperature, and fluid property data.

  ◦ The acquired information is used to determine reservoir capabilities and important decisions such as production methods, well production equipment, and field development drilling are made from the interpretation of well test results.

Well testing on board of semisub rig.

Petroleum Production

Petroleum production is the recovery of crude oil and, often, associated natural gas from Earth.

Petroleum is a naturally occurring hydrocarbon material that is believed to have formed from animal and vegetable debris in deep sedimentary beds. The petroleum, being less dense than the surrounding water, was expelled from the source beds and migrated upward through porous rock such as sandstone and some limestone until it was finally blocked by nonporous rock such as shale or dense limestone. In this way, petroleum deposits came to be trapped by geologic features caused by the folding, faulting, and erosion of Earth's crust.

The Trans-Alaska Pipeline running parallel to a highway north of Fairbanks.

Petroleum may exist in gaseous, liquid, or near-solid phases either alone or in combination. The liquid phase is commonly called crude oil, while the more-solid phase may be called bitumen, tar, pitch, or asphalt. When these phases occur together, gas usually overlies the liquid, and the liquid overlies the more-solid phase. Occasionally, petroleum deposits elevated during the formation of mountain ranges have been exposed by erosion to form tar deposits. Some of these deposits have been known and exploited throughout recorded history. Other near-surface deposits of liquid petroleum seep slowly to the surface through natural fissures in the overlying rock. Accumulations from these seeps, called rock oil, were used commercially in the 19th century to make lamp oil by simple distillation. The vast majority of petroleum deposits, however, lie trapped in the pores of natural rock at depths from 150 to 7,600 metres (500 to 25,000 feet) below the surface of the ground. As a general rule, the deeper deposits have higher internal pressures and contain greater quantities of gaseous hydrocarbons.

When it was discovered in the 19th century that rock oil would yield a distilled product (kerosene) suitable for lanterns, new sources of rock oil were eagerly sought. It is now generally agreed that the first well drilled specifically to find oil was that of Edwin Laurentine Drake in Titusville, Pennsylvania, U.S., in 1859. The success of this well, drilled close to an oil seep, prompted further drilling in the same vicinity and soon led to similar exploration elsewhere. By the end of the century, the growing demand for petroleum products resulted in the drilling of oil wells in other states and countries. In 1900, crude oil production worldwide was nearly 150 million barrels. Half of this total was produced in Russia, and most (80 percent) of the rest was produced in the United States.

First oil well in the United States, built in 1859.

First oil wells pumping in the United States.

From the discovery of the first oil well in 1859 until 1870, the annual production of oil in the United States increased from about two thousand barrels to nearly ten million. In 1870 John D. Rockefeller formed the Standard Oil Company, which eventually controlled virtually the entire industry. The Standard, while ruthless in business methods, was largely responsible for the rapid growth of refining and distribution techniques.

The advent and growth of automobile usage in the second decade of the 20th century created a great demand for petroleum products. Annual production surpassed one billion barrels in 1925 and two billion barrels in 1940. By the last decade of the 20th century, there were almost one million wells in more than 100 countries producing more than 20 billion barrels per year. By the end of the second decade of the 21st century, petroleum production had risen to nearly 34 billion barrels per year, of which an increasing share was supported by ultradeepwater drilling and unconventional crude production (in which petroleum is extracted from shales, tar sands, or bitumen or is recovered by other methods that differ from conventional drilling). Petroleum is produced on every continent except Antarctica, which is protected from petroleum exploration by an environmental protocol to the Antarctic Treaty until 2048.

Prospecting and Exploration

Drake's original well was drilled close to a known surface seepage of crude oil. For years such seepages were the only reliable indicators of the presence of underground oil and gas. However, as demand grew, new methods were devised for evaluating the potential of underground rock formations. Today, exploring for oil requires integration of information collected from seismic surveys, geologic framing, geochemistry, petrophysics, geographic information systems (GIS) data gathering, geostatistics, drilling, reservoir engineering, and other surface and subsurface investigative techniques. Geophysical exploration including seismic analysis is the primary method of exploring for petroleum. Gravity and magnetic fieldmethods are also historically reliable evaluation methods carrying over into more complex and challenging exploration environments, such as sub-salt structures and deep water. Beginning with GIS, gravity, magnetic, and seismic surveys allow geoscientists to efficiently focus the search for target assets to explore, thus lowering the risks associated with exploration drilling.

Natural oil seep.

There are three major types of exploration methods: (1) surface methods, such as geologic feature mapping, enabled by GIS, (2) area surveys of gravity and magnetic fields, and (3) seismographic methods. These methods indicate the presence or absence of subsurface features that are favourable for petroleum accumulations. There is still no way to predict the presence of productive underground oil deposits with 100 percent accuracy.

Surface Methods

Crude oil seeps sometimes appear as a tarlike deposit in a low area—such as the oil springs at Baku, Azerbaijan, on the Caspian Sea, described by Marco Polo. More often they occur as a thin skim of oil on small creeks that pass through an area. This latter phenomenon was responsible for the naming of Oil Creek in Pennsylvania, where Drake's well was drilled. Seeps of natural gas usually cannot be seen, although instruments can detect natural gas concentrations in air as low as 1 part in 100,000. Similar instruments have been used to test for traces of gas in seawater. These geochemical surface prospecting methods are not applicable to the large majority of petroleum reservoirs, which do not have leakage to the surface.

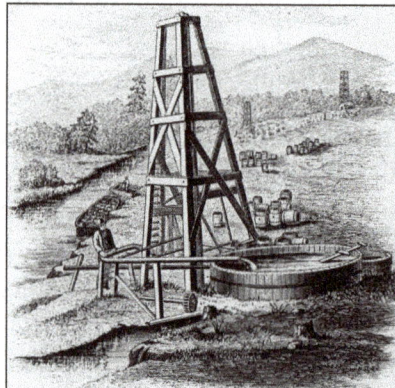

Oil wells on Oil Creek, near the Allegheny River in Pennsylvania, U.S.

Another method is based on surface indications of likely underground rock formations. In some cases, subsurface folds and faults in rock formations are repeated in the surface features. The presence of underground salt domes, for example, may be indicated by a low bulge in an otherwise flat ground surface. Uplifting and faulting in the rock formations surrounding these domes often result in oil and gas accumulations.

Gravity and Magnetic Surveys

Although gravity at Earth's surface is very nearly constant, it is slightly greater where dense rock formations lie close to the surface. Gravitational force, therefore, increases over the tops of anticlinal (arch-shaped) folds and decreases over the tops of salt domes. Very small differences in gravitational force can be measured by a sensitive instrument known as the gravimeter. Measurements are made on a precise grid over a large area, and the results are mapped and interpreted to reflect the presence of potential oil- or gas-bearing formations.

Magnetic surveys make use of the magnetic properties of certain types of rock that, when close to the surface, affect Earth's normal magnetic field. Again, sensitive instruments are used to map anomalies over large areas. Surveys are often carried out from aircraft over land areas and from oceangoing vessels over continental shelves. A similar method, called magnetotellurics (MT), measures the natural electromagnetic field at Earth's surface. The different electrical resistivities of rock formations cause anomalies that, when mapped, are interpreted to reflect underground geologic features. MT is becoming a more cost-effective filter to identify a petroleum play (a set of oil fields or petroleum deposits with similar geologic characteristics) before more costly and time-intensive seismic surveying is conducted. MT is sensitive to what is contained within Earth's stratographic layers. Crystalline rocks such as subsalts (that is, salts whose bases are not fully neutralized by acid) tend to be very resistive to electromagnetic waves, whereas porous rocks are usually conductive because of the seawater and brines contained within them. Petroleum geologists look to anomalies such as salt domes as indicators of potential stratigraphic traps for petroleum.

Seismographic Methods

The survey methods described above can show the presence of large geologic anomalies such as anticlines (arch-shaped folds in subterranean layers of rock), fault blocks (sections of rock layers separated by a fracture or break), and salt domes, even though there may not be surface indications of their presence. However, they cannot be relied upon to find smaller and less obvious traps and unconformities (gaps) in the stratigraphic arrangement of rock layers that may harbour petroleum reservoirs. These can be detected and located by seismic surveying, which makes use of the sound-transmitting and sound-reflecting properties of underground rock formations. Seismic waves travel at different velocities through different types of rock formations and are reflected by the interfaces between different types of rocks. The sound-wave source is usually a small explosion in a shallow drilled hole. Microphones are placed at various distances and directions from the explosive point to pick up and record the transmitted and reflected sound-wave arrivals. The procedure is repeated at intervals over a wide area. An experienced seismologist can then interpret the collected records to map the underground formation contours.

Offshore and land-based seismic data collection varies primarily by method of setup. For offshore seismic surveys, one of the most critical components of petroleum exploration is knowing where the ship and receivers are at all times, which is facilitated by relaying global positioning system (GPS) readings in real time from satellites to GPS reference and monitoring stations and then to the ship. Readings in real time have become part of the process of seismic sound-wave capture, data processing, and analysis.

Offshore Seismic Acquisition

Sound is often generated by air guns, and the sonic returns produce images of the shear waves in the water and subsurface. Towed hydrophone arrays (also called hydrophone streamers) detect the sound waves that return to the surface through the water and sub-seafloor strata. Reflected sound is recorded for the elapsed travel time and the strength of the returning sound waves. Successful seismic processing requires an accurate reading of the returning sound waves, taking into account how the various gaseous, liquid, and solid media the sound waves travel through affect the progress of the sound waves.

Two-dimensional (2-D) seismic data are collected from each ship that tows a single hydrophone streamer. The results display as a single vertical plane or in cross section that appears to slice into the subsurface beneath the seismic line. Interpretation outside the plane is not possible with two-dimensional surveys; however, it is possible with three-dimensional (3-D) ones. The utility of 2-D surveys is in general petroleum exploration or frontier exploration. In this work, broad reconnaissance is often required to identify focus areas for follow-up analysis using 3-D techniques.

Seismic data collection in three dimensions employs one or more towed hydrophone streamers. The arrays are oriented so that they are towed in a linear fashion, such as in a "rake" pattern (where several lines are towed in parallel), to cover the area of interest. The results display as a three-dimensional cube in the computer environment. The cube can be sliced and rotated by using various software for processing and analysis. In addition to better resolution, 3-D processed data produce spatially continuous results, which help to reduce the uncertainty in marking the boundaries of a deposit, especially in areas where the geology is structurally complex or in cases where the deposits are small and thus easily overlooked. Going one step further, two 3-D data sets from different periods of time can be combined to show volumetric or other changes in oil, water, or gas in a reservoir, essentially producing a four-dimensional seismic survey with time being the fourth dimension.

On rare occasions and at shallower depths, receivers can be physically placed on the seafloor. Cost and time factor into this method of data acquisition, but this technique may be preferred when towing hydrophone streamers would be problematic, such as in shipping lanes or near rigid offshore structures or commercial fishing operations.

Land-based Seismic Acquisition

Onshore seismic data have been acquired by using explosions of dynamite to produce sound waves as well as by using the more environmentally sensitive vibroseis system (a vibrating mechanism that creates seismic waves by striking Earth's surface). Dynamite is used away from populated areas where detonation can be secured in plugged shot holes below the surface layer. This method is preferred to vibroseis, since it gives sharp, clean sound waves. However, more exploration efforts are shifting to vibroseis, which incorporates trucks capable of pounding the surface with up to nearly 32 metric tons (approximately 35 tons) of force. Surface pounding creates vibrations that produce seismic waves, which generate data similar to those of offshore recordings.

Processing and Visualization

Processing onshore and offshore seismic data is a complex effort. It begins with filtering massive amounts of data for output and background noise during seismic capture. The filtered data are then formally processed—which involves the deconvolution (or sharpening) of the "squiggly lines" correlating to rock layers, the gathering and summing of stacked seismic traces (digital curves or returns from seismic surveys) from the same reflecting points, the focusing of seismic traces to fill in the gaps or smoothed-over areas that lack trace data, and the manipulation of the output to give the true, original positions of the trace data.

With more computer power, integrating seismic processing and its analysis with other activities that define the geologic context of the scanned area has become a routine task in the 21st century. Visualizing the collected data for purposes of exploration and production began with the introduction of interpretation workstations in the early 1980s, and technology designed to help researchers interpret volumetric pixels (3-D pixels, or "voxels") became available in the early 1990s. Advances in graphics, high-performance computing, and artificial intelligence supported and expanded data visualization tasks. By the early 21st century, data visualization in oil exploration and production was integrating these advances while also illustrating to the geoscientist and engineer the increasing uncertainty and complexity of the available information.

Visualization setups incorporate seismic data alongside well logs (physical data profiles taken in or around a well or borehole) or petrophysical data taken from cores (cylindrical rock samples). The visualization setups typically house complex data and processes to convert statistical data into graphical analyses in multiple sizes or shapes. The data display can vary widely, with front or rear projections from spherical, cylindrical, conical, or flat screens; screen sizes range from small computer monitors to large-scale dome configurations. The key results from using visualization are simulations depicting interactive reservoirs of flowing oil and trials designed to test uncertain geological features at or below the resolution of seismic data.

Oil Well

Drilling

1. Cable Tooling

Early oil wells were drilled with impact-type tools in a method called cable-tool drilling. A weighted chisel-shaped bit was suspended from a cable to a lever at the surface, where an up-and-down motion of the lever caused the bit to chip away the rock at the bottom of the hole. The drilling had to be halted periodically to allow loose rock chips and liquids to be removed with a collecting device attached to the cable. At these times the chipping tip of the bit was sharpened, or "dressed" by the tool dresser. The borehole had to be free of liquids during the drilling so that the bit could remove rock effectively. This dry condition of the hole allowed oil and gas to flow to the surface when the bit penetrated a producing formation, thus creating the image of a "gusher" as a successful oil well. Often a large amount of oil was wasted before the well could be capped and brought under control.

2. The Rotary Drill

During the mid- to late 20th century, rotary drilling became the preferred penetration method for oil and gas wells. In this method a special tool, the drill bit, rotates while bearing down on the bottom of the well, thus gouging and chipping its way downward. Probably the greatest advantage of rotary drilling over cable tooling is that the well bore is kept full of liquid during drilling. A weighted fluid (drilling mud) is circulated through the well bore to serve two important purposes. By its hydrostatic pressure, it prevents entry of the formation fluids into the well, thereby preventing blowouts and gushers (uncontrolled oil releases). In addition, the drilling mud carries the crushed rock to the surface, so that drilling is continuous until the bit wears out.

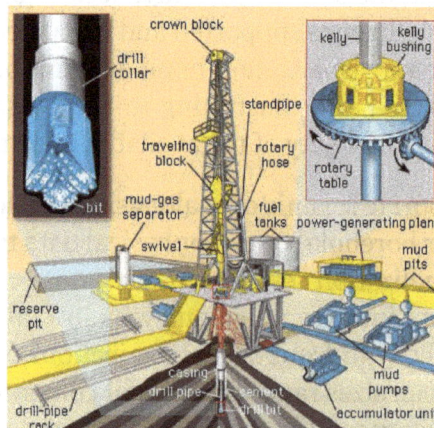

A land-based rotary drilling rig.

Rotary drilling techniques have enabled wells to be drilled to depths of more than 9,000 metres (30,000 feet). Formations having fluid pressures greater than 1,400 kg per square cm (20,000 pounds per square inch) and temperatures greater than 250 °C (480 °F) have been successfully penetrated. Additionally, improvements to rotary drilling techniques have reduced the time it takes to drill long distances. A powered rotary steerable system (RSS) that can be controlled and monitored remotely has become the preferred drilling technology for extended-reach drilling (ERD) and deepwater projects. In some cases, onshore well projects that would have taken 35 days to drill in 2007 could be finished in only 20 days 10 years later by using the RSS. Offshore, one of the world's deepest wells in the Chayvo oil field, off the northeastern corner of Sakhalin Island in Russia, was drilled by Exxon Neftegas Ltd. using its "fast drilling" process. The Z-44 well, drilled in 2012, is 12,345 metres (about 40,500 feet) deep.

A common tricone oil-drill bit with three steel cones rotating on bearings.

3. The Drill Pipe

The drill bit is connected to the surface equipment through the drill pipe, a heavy-walled tube through which the drilling mud is fed to the bottom of the borehole. In most cases, the drill pipe also transmits the rotary motion to the bit from a turntable at the surface. The top piece of the drill pipe is a tube of square (or occasionally six- or eight-sided) cross section called the kelly. The kelly passes through a similarly shaped hole in the turntable. At the bottom end of the drill pipe are extra-heavy sections called drill collars, which serve to concentrate the weight on the rotating bit. In order to help maintain a vertical well bore, the drill pipe above the collars is usually kept in tension. The drilling mud leaves the drill pipe through the bit in such a way that it scours the loose rock from the bottom and carries it to the surface. Drilling mud is carefully formulated to assure the correct weight and viscosity properties for the required tasks. After screening to remove the rock chips, the mud is held in open pits or metal tanks to be recirculated through the well. The mud is picked up by piston pumps and forced through a swivel joint at the top of the kelly.

Three oil-rig roughnecks pulling drill pipe out of an oil well.

4. The Derrick

The hoisting equipment that is used to raise and lower the drill pipe, along with the machinery for rotating the pipe, is contained in the tall derrick that is characteristic of rotary drilling rigs. While early derricks were constructed at the drilling site, modern rigs can be moved from one site to the next. The drill bit wears out quickly and requires frequent replacement, often once a day. This makes it necessary to pull the entire drill string (the column of drill pipe) from the well and stand all the joints of the drill pipe vertically at one side of the derrick. Joints are usually 9 metres (29.5 feet) long. While the bit is being changed, sections of two or three joints are separated and stacked. Drilling mud is left in the hole during this time to prevent excessive flow of fluids into the well.

Workers on an oil rig.

5. Casing

Modern wells are not drilled to their total depth in a continuous process. Drilling may be stopped for logging and testing, and it may also be stopped to run (insert) casing and cement it to the outer circumference of the borehole. (Casing is steel pipe that is intended to prevent any transfer of fluids between the borehole and the surrounding formations). Since the drill bit must pass through any installed casing in order to continue drilling, the borehole below each string of casing is smaller than the borehole above. In very deep wells, as many as five intermediate strings of progressively smaller-diameter casing may be used during the drilling process.

6. The Turbodrill

One variation in rotary drilling employs a fluid-powered turbine at the bottom of the borehole to produce the rotary motion of the bit. Known as the turbodrill, this instrument is about nine metres long and is made up of four major parts: the upper bearing, the turbine, the lower bearing, and the drill bit. The upper bearing is attached to the drill pipe, which either does not rotate or rotates at a slow rate (6 to 8 revolutions per minute). The drill bit, meanwhile, rotates at a much faster rate (500 to 1,000 revolutions per minute) than in conventional rotary drilling. The power source for the turbodrill is the mud pump, which forces mud through the drill pipe to the turbine. The mud is diverted onto the rotors of the turbine, turning the lower bearing and the drill bit. The mud then passes through the drill bit to scour the hole and carry chips to the surface.

The turbodrill is capable of very fast drilling in harsh environments, including high-temperature and high-pressure rock formations. Periodic technological improvements have included longer-wearing bits and bearings. Turbodrills were originally developed and widely used in Russia and Central Asia. Given their capabilities for extended reach and drilling in difficult rock formations, turbodrill applications expanded into formerly inaccessible regions on land and offshore. Turbodrills with diamond-impregnated drill bits became the choice for hard, abrasive rock formations. The high rotating speeds exceeded more than 1,000 revolutions per minute, which facilitated faster rates of penetration (ROPs) during drilling operations.

7. Directional Drilling

Frequently, a drilling platform and derrick cannot be located directly above the spot where a well should penetrate the formation (if, for example, a petroleum reservoir lies under a lake, town, or harbour). In such cases, the surface equipment must be offset and the well bore drilled at an angle that will intersect the underground formation at the desired place. This is done by drilling the well vertically to start and then angling it at a depth that depends on the relative position of the target. Since the nearly inflexible drill pipe must be able to move and rotate through the entire depth, the angle of the borehole can be changed only a few degrees per tens of feet at any one time. In order to achieve a large deviation angle, therefore, a number of small deviations must be made. The borehole, in effect, ends up making a large arc to reach its objective. The original tool for "kicking off" such a well was a mechanical device called the whipstock. This consisted of an inclined plane on the bottom of the drill pipe that was oriented in the direction the well was intended to take. The drill bit was thereby forced to move off in the proper direction. A more recent technique makes use of steerable motor assemblies containing positive-displacement motors (PDMs) with adjustable bent-housing mud motors. The bent housing misaligns the bit face away from the line of the drill string, which causes the bit to change the direction of the hole being drilled. PDM bent-housing motor assemblies are most commonly used to "sidetrack" out of existing casing. (Sidetracking is drilling horizontal lateral lines out from existing well bores [drill holes]). In mature fields where engineers and drilling staff target smaller deposits of oil that were bypassed previously, it is not uncommon to use existing well bores to develop the bypassed zones. In order to accomplish this, a drill string is prepared to isolate the other producing zones. Later, a casing whipstock is used to mill (or grind) through the existing casing. The PDM bent-housing motor assembly is then run into the cased well to divert the trajectory of the drill so that the apparatus can point toward the targeted deposit.

As more-demanding formations are encountered—such as in ultradeep, high-pressure, high-temperature, abrasive rock and shales—wear and tear on the mud motors and bits causes frequent "trips." (Trips involve pulling worn-out mechanical bits and motors from the well, attaching replacements, and reentering the well to continue drilling). To answer these challenges, modern technologies incorporate an RSS capable of drilling vertical, curved, and horizontal sections in one trip. During rotary steering drilling, a surface monitoring system sends steering control commands to the downhole steering tools in a closed-loop control system. In essence, two-way communication between the surface and the downhole portions of the equipment improves the drilling rate of penetration (ROP). The surface command transmits changes in the drilling fluid pressure and flow rate in the drilling pipe. Pulse signals of drilling fluid pressure with different pulse widths are generated by adjusting the timing of the pulse valve, which releases the drilling fluid into the pipe.

Further advances to the RSS include electronically wired drill pipe that is intended to speed communication from the surface to the bit. This technology has matured to the point where it coordinates with logging-while-drilling (LWD) systems. It also provides faster data transfer than pulsed signaling techniques and continuous data in real time from the bottom hole assembly. The safety advantages, however, perhaps trump the increases in the rate of information transfer. Knowing the downhole temperature and pressure data in real time can give the operator advance notice of changing formation conditions, which allows the operator more control over the well.

Smart field technologies, such as directional drilling techniques, have rejuvenated older fields by accessing deposits that were bypassed in the past in favour of more easily extractable plays. Directional drilling techniques have advanced to the point where well bores can end in horizontal sections extending into previously inaccessible areas of a reservoir. Also, multiple deposits can be accessed through extended-reach drilling by a number of boreholes fanning out from a single surface structure or from various points along a vertical borehole. Technology has allowed once noncommercial resources, such as those found in harsh or relatively inaccessible geologic formations, to become developable reserves.

Offshore Platforms

Shallow Water

Many petroleum reservoirs are found in places where normal land-based drilling rigs cannot be used. In inland waters or wetland areas, a drilling platform and other drilling equipment may be mounted on a barge, which can be floated into position and then made to rest on the seafloor. The actual drilling platform can be raised above the water on masts if necessary. Drilling and other operations on the well make use of an opening through the barge hull. This type of rig is generally restricted to water depths of 15 metres (50 feet) or less.

Oil derricks in the Caspian Sea near Baku.

In shallow Arctic waters where drifting ice is a hazard for fixed platforms, artificial islands have been constructed of rock or gravel. Onshore in Arctic areas, permafrost makes drilling difficult because melting around and under the drill site makes the ground unstable. There too, artificial islands are built up with rock or gravel.

Away from the nearshore zone, shallow offshore drilling takes place in less than 152 metres (500 feet) of water, which permits the use of fixed platforms with concrete or metal legs planted into the seafloor. Control equipment resides at the surface, on the platform with the wellhead positioned on the seafloor. When the water depth is less than 457 metres (1,500 feet), divers can easily reach the wellhead to perform routine maintenance as required, which makes shallow offshore drilling one of the safest methods of offshore production.

Deep and Ultradeep Water

In deeper, more open waters up to 5,000 feet (1,524 metres) deep over continental shelves, drilling is done from free-floating platforms or from platforms made to rest on the bottom. Floating rigs are most often used for exploratory drilling and drilling in waters deeper than 3,000 feet (914 metres), while bottom-resting platforms are usually associated with the drilling of wells in an established field or in waters shallower than 3,000 feet. One type of floating rig is the drill ship, which is used almost exclusively for exploration drilling before commitments to offshore drilling and production are made. This is an oceangoing vessel with a derrick mounted in the middle, over an opening for the drilling operation. Such ships were originally held in position by six or more anchors, although some vessels were capable of precise maneuvering with directional thrust propellers. Even so, these drill ships roll and pitch from wave action, making the drilling difficult. At present, dynamic positioning gear systems are affixed to drill ships, which permit operations in heavy seas or other severe conditions.

The Jack Ryan, a drill ship capable of exploring for oil in water 3,000 metres (10,000 feet) deep.

Floating deepwater drilling and petroleum production methods vary, but they all involve the use of fixed (anchored) systems, which may be put in place once drilling is complete and the drilling rig demobilized. Additional production is established by a direct connection with the production platform or by connecting risers between the subsea wellheads and the production platform. The Seastar floating system operates in waters up to 3,500 feet (1,067 metres) deep. It is essentially a small-scale tension-leg platform system that allows for side-to-side movement but minimizes up-and-down movement. Given the vertical tension, production is tied back to "dry" wellheads (on the surface) or to "trees" (structures made up of valves and flow controls) on the platform that are similar to those of the fixed systems.

Jack-up oil platform: A jack-up rig drilling for oil in the Caspian Sea.

Semisubmersible deepwater production platforms are more stable. Their buoyancy is provided by a hull that is entirely underwater, while the operational platform is held well above the surface on supports. Normal wave action affects such platforms very little. These platforms are commonly kept in place during drilling by cables fastened to the seafloor. In some cases the platform is pulled down on the cables so that its buoyancy creates a tension that holds it firmly in place. Semisubmersible platforms can operate in ultradeep water—that is, in waters more than 3,050 metres (10,000 feet) deep. They are capable of drilling to depths of more than 12,200 metres (approximately 40,000 feet).

Drilling platforms capable of ultradeepwater production—that is, beyond 1,830–2,130 metres (approximately 6,000–7,000 feet) deep—include tension-leg systems and floating production systems (FPS), which can move up and down in response to ocean conditions as semisubmersibles perform. The option to produce from wet (submerged) or dry trees is considered with respect to existing infrastructure, such as regional subsea pipelines. Without such infrastructure, wet trees are used and petroleum is exported to a nearby FPS. A more versatile ultradeepwater system is the spar type, which can perform in waters nearly 3,700 metres (approximately 12,000 feet) deep. Spar systems are moored to the seabed and designed in three configurations: (1) a conventional one-piece cylindrical hull, (2) a truss spar configuration, where the midsection is composed of truss elements connecting an upper, buoyant hull (called a hard tank) with a bottom element (soft tank) containing permanent ballast, and (3) a cell spar, which is built from multiple vertical cylinders. In the cell spar configuration, none of the cylinders reach the seabed, but all are tethered to the seabed by mooring lines.

Fixed platforms, which rest on the seafloor, are very stable, although they cannot be used to drill in waters as deep as those in which floating platforms can be used. The most popular type of fixed platform is called a jack-up rig. This is a floating (but not self-propelled) platform with legs that can be lifted high off the seafloor while the platform is towed to the drilling site. There the legs are cranked downward by a rack-and-pinion gearing system until they encounter the seafloor and actually raise the platform 10 to 20 metres (33 to 66 feet) above the surface. The bottoms of the legs are usually fastened to the seafloor with pilings. Other types of bottom-setting platforms, such as the compliant tower, may rest on flexible steel or concrete bases that are constructed onshore to the correct height. After such a platform is towed to the drilling site, flotation tanks built into the base are flooded, and the base sinks to the ocean floor. Storage tanks for produced oil may be built into the underwater base section.

Three types of offshore drilling platforms.

For both fixed rigs and floating rigs, the drill pipe must transmit both rotary power and drilling mud to the bit; in addition, the mud must be returned to the platform for recirculation. In order to accomplish these functions through seawater, an outer casing, called a riser, must extend from the seafloor to the platform. Also, a guidance system (usually consisting of cables fastened to the seafloor) must be in place to allow equipment and tools from the surface to enter the well bore. In the case of a floating platform, there will always be some motion of the platform relative to the seafloor, so this equipment must be both flexible and extensible. A guidance system will be especially necessary if the well is to be put into production after the drilling platform is moved away.

The Thunder Horse, a semisubmersible oil production platform, constructed to operate several wells in waters more than 1,500 metres (5,000 feet) deep in the Gulf of Mexico.

Using divers to maintain subsea systems is not as feasible in deep waters as in shallow waters. Instead, an intricate system of options has been developed to distribute risks away from any one subsea source, such as a wet tree. Smart well control and connection systems assist from the seafloor in directing subsea manifolds, pipelines, risers, and umbilicals prior to oil being lifted to the surface. Subsea manifolds direct the subsea systems by connecting wells to export pipelines and risers and onward to receiving tankers, pipelines, or other facilities. They direct produced oil to flowlines coincidental to distributing injected water, gas, or chemicals.

The reliance on divers in subsea operations began to fade in the 1970s, when the first unmanned vehicles or remotely operated vehicles (ROVs) were adapted from space technologies. ROVs became essential in the development of deepwater reserves. Robotics technology, which was developed primarily for the ROV industry, has been adapted for a wide range of subsea applications.

Formation Evaluation

Advances in technology have occurred in well logging and the evaluation of geological formations more than in any other area of petroleum production. Historically, after a borehole penetrated a potential productive zone, the formations were tested to determine their nature and the degree to which completion procedures (the series of steps that convert a drilling well into a producing well) should be conducted. The first evaluation was usually made using well logging methods. The logging tool was lowered into the well by a steel cable and was pulled past the formations while response signals were relayed to the surface for observation and recording. Often these tools made use of the differences in electrical conductivities of rock, water, and petroleum to detect possible oil or gas accumulations. Other logging tools used differences in radioactivity, neutron absorption, and acoustic wave absorption. Well log analysts could use the recorded signals to determine potential producing formations and their exact depth. Only a production, or "formation," test, however, could establish the potential productivity.

The production test that was historically employed was the drill stem test, in which a testing tool was attached to the bottom of the drill pipe and was lowered to a point opposite the formation to be tested. The tool was equipped with expandable seals for isolating the formation from the rest of the borehole, and the drill pipe was emptied of mud so that formation fluid could enter. When enough time had passed, the openings into the tool were closed and the drill pipe was brought to the surface so that its contents could be measured. The amounts of oil and gas that flowed into the drill pipe during the test and the recorded pressures were used to judge the production potential of the formation.

With advances in measurement-while-drilling (MWD) technologies, independent well logging and geological formation evaluation runs became more efficient and more accurate. Other improvements in what has become known as smart field technologies included a widening range of tool sizes and deployment options that enable drilling, logging, and formation evaluation into smaller boreholes simultaneously. Formation measurement techniques that employ logging-while-drilling (LWD) equipment include gamma ray logging, resistivity measurement, density and neutron porosity logging, sonic logging, pressure testing, fluid sampling, and borehole diameter measurements using calipers. LWD applications include flexible logging systems for horizontal wells in shale plays with curvatures as sharp as 68° per 100 feet. Another example of an improvement in smart field technologies is use of rotary steerable systems in deep waters, where advanced LWD is vastly reducing the evaluation time of geological formations, especially in deciding whether to complete or abandon a well. Reduced decision times have led to an increase in the safety of drilling, and completion operations have become much improved, as the open hole is cased or plugged and abandoned that much sooner. With traditional wireline logs, reports of findings may not be available for days or weeks. In comparison, LWD coupled with RSS is controlled by the drill's ROP. The formation evaluation sample rate combined with the ROP determine the eventual number of measurements per drilled foot that will be recorded on the log. The faster the ROP, the faster the sample rate and its recording onto the well log sent to the surface operator for analysis and decision making.

Well Completion

Production Tubing

If preliminary tests show that one or more of the formations penetrated by a borehole will be commercially productive, the well must be prepared for the continuous production of oil or gas. First, the casing is completed to the bottom of the well. Cement is then forced into the annulus between the casing and the borehole wall to prevent fluid movement between formations. As mentioned earlier, this casing may be made up of progressively smaller-diameter tubing, so that the casing diameter at the bottom of the well may range from 10 to 30 cm (4 to 12 inches). After the casing is in place, a string of production tubing 5 to 10 cm (2 to 4 inches) in diameter is extended from the surface to the productive formation. Expandable packing devices are placed on the tubing to seal the annulus that lies between the casing and the production tubing within the producing formation from the annulus that lies within the remainder of the well. If a lifting device is needed to bring the oil to the surface, it is generally placed at the bottom of the production tubing. If several producing formations are penetrated by a single well, as many as four production strings may be hung. However, as deeper formations are targeted, conventional completion practices often produce diminishing returns.

Perforating and Fracturing

Since the casing is sealed with cement against the productive formation, openings must be made in the casing wall and cement to allow formation fluid to enter the well. A perforator tool is lowered through the tubing on a wire line. When it is in the correct position, bullets are fired or explosive charges are set off to create an open path between the formation and the production string. If the formation is quite productive, these perforations (usually about 30 cm, or 12 inches, apart) will be sufficient to create a flow of fluid into the well. If not, an inert fluid may be injected into the formation at pressure high enough to cause fracturing of the rock around the well and thus open more flow passages for the petroleum.

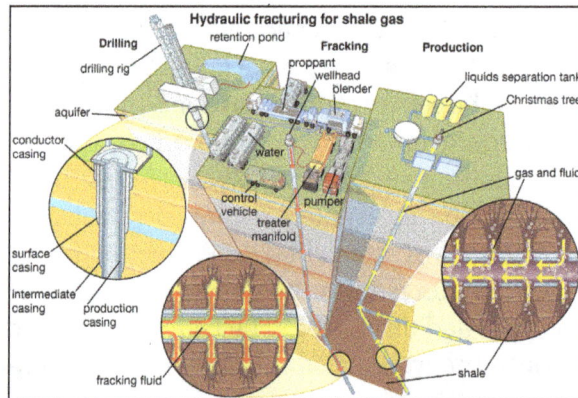

Three steps in the extraction of shale gas: Drilling a borehole into the shale formation and lining it with pipe casing; fracking, or fracturing, the shale by injecting fluid under pressure; and producing gas that flows up the borehole, frequently accompanied by liquids.

Tight oil formations are typical candidates for hydraulic fracturing (fracking), given their characteristically low permeability and low porosity. During fracturing, water, which may be accompanied by sand, and less than 1 percent household chemicals, which serve as additives, are pumped into the reservoir at high pressure and at a high rate, causing a fracture to open. Sand, which served as the propping agent (or "proppant"), is mixed with the fracturing fluids to keep the fracture open. When the induced pressure is released, the water flows back from the well with the proppant remaining to prop up the reservoir rock spaces. The hydraulic fracturing process creates network of interconnected fissures in the formation, which makes the formation more permeable for oil, so that it can be accessed from beyond the near-well bore area.

In early wells, nitroglycerin was exploded in the uncased well bore for the same purpose. An acid that can dissolve portions of the rock is sometimes used in a similar manner.

Surface Valves

When the subsurface equipment is in place, a network of valves, referred to as a Christmas tree, is installed at the top of the well. The valves regulate flow from the well and allow tools for subsurface work to be lowered through the tubing on a wire line. Christmas trees may be very simple, as in those found on low-pressure wells that must be pumped, or they may be very complex, as on high-pressure flowing wells with multiple producing strings.

A worker operating a "Christmas tree," a structure of valves for regulating flow at the surface of an oil well.

Recovery of Oil and Gas

Primary Recovery: Natural Drive and Artificial Lift

Petroleum reservoirs usually start with a formation pressure high enough to force crude oil into the well and sometimes to the surface through the tubing. However, since production is invariably accompanied by a decline in reservoir pressure, "primary recovery" through natural drive soon comes to an end. In addition, many oil reservoirs enter production with a formation pressure high enough to push the oil into the well but not up to the surface through the tubing. In these cases, some means of "artificial lift" must be installed. The most common installation uses a pump at the bottom of the production tubing that is operated by a motor and a "walking beam" (an arm that rises and falls like a seesaw) on the surface. A string of solid metal "sucker rods" connects the walking beam to the piston of the pump. Another method, called gas lift, uses gas bubbles to lower the density of the oil, allowing the reservoir pressure to push it to the surface. Usually, the gas is injected down the annulus between the casing and the production tubing and through a special valve at the bottom of the tubing. In a third type of artificial lift, produced oil is forced down the well at high pressure to operate a pump at the bottom of the well.

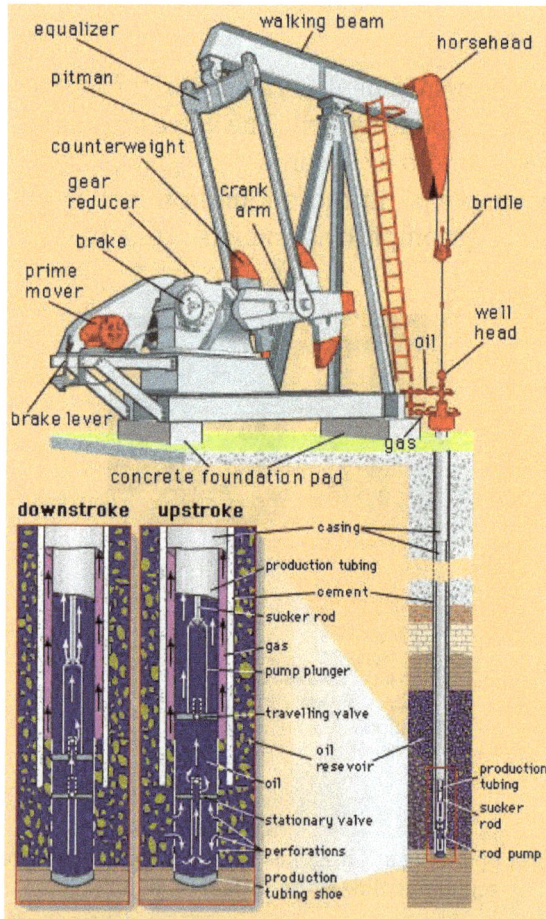

The "artificial lift" of petroleum with a beam-pumping unit.

With hydraulic lift systems, crude oil or water is taken from a storage tank and fed to the surface pump. The pressurized fluid is distributed to one or more wellheads. For cost-effectiveness, these artificial lift systems are configured to supply multiple wellheads in a pad arrangement, a configuration where several wells are drilled near each other. As the pressurized fluid passes into the wellhead and into the downhold pump, a piston pump engages that pushes the produced oil to the surface. Hydraulic submersible pumps create an advantage for low-volume producing reservoirs and low-pressure systems.

A "walking beam" operating an oil well pump.

Conversely, electrical submersible pumps (ESPs) and downhole oil water separators (DOWS) have improved primary production well life for high-volume wells. ESPs are configured to use centrifugal force to artificially lift oil to the surface from either vertical or horizontal wells. ESPs are useful because they can lift massive volumes of oil. In older fields, as more water is produced, ESPs are preferred for "pumping off" the well to permit maximum oil production. DOWS provide a method to eliminate the water handling and disposal risks associated with primary oil production, by separating oil and gas from produced water at the bottom of the well. Oil and gas are later pumped to the surface while water associated with the process is reinjected into a disposal zone below the surface.

With the artificial lift methods described above, oil may be produced as long as there is enough nearby reservoir pressure to create flow into the well bore. Inevitably, however, a point is reached at which commercial quantities no longer flow into the well. In most cases, less than one-third of the oil originally present can be produced by naturally occurring reservoir pressure alone. In some cases (e.g., where the oil is quite viscous and at shallow depths), primary production is not economically possible at all.

Secondary Recovery: Injection of Gas or Water

When a large part of the crude oil in a reservoir cannot be recovered by primary means, a method for supplying extra energy must be found. Most reservoirs have some gas in a miscible state, similar to that of a soda bottled under pressure before the gas bubbles are released when the cap is opened. As the reservoir produces under primary conditions, the solution gas escapes, which lowers the pressure of the reservoir. A "secondary recovery" is required to reenergize or "pressure up" the reservoir. This is accomplished by injecting gas or water into the reservoir to replace produced fluids and thus maintain or increase the reservoir pressure. When gas alone is injected, it is usually put into the top of the reservoir, where petroleum gases normally collect to form a gas cap. Gas injection can be a very effective recovery method in reservoirs where the oil is able to flow freely to the bottom by gravity. When this gravity segregation does not occur, however, other means must be sought.

An even more widely practiced secondary recovery method is waterflooding. After being treated to remove any material that might interfere with its movement in the reservoir, water is injected through some of the wells in an oil field. It then moves through the formation, pushing oil toward the remaining production wells. The wells to be used for injecting water are usually located in a pattern that will best push oil toward the production wells. Water injection often increases oil recovery to twice that expected from primary means alone. Some oil reservoirs (the East Texas field, for example) are connected to large, active water reservoirs, or aquifers, in the same formation. In such cases it is necessary only to reinject water into the aquifer in order to help maintain reservoir pressure.

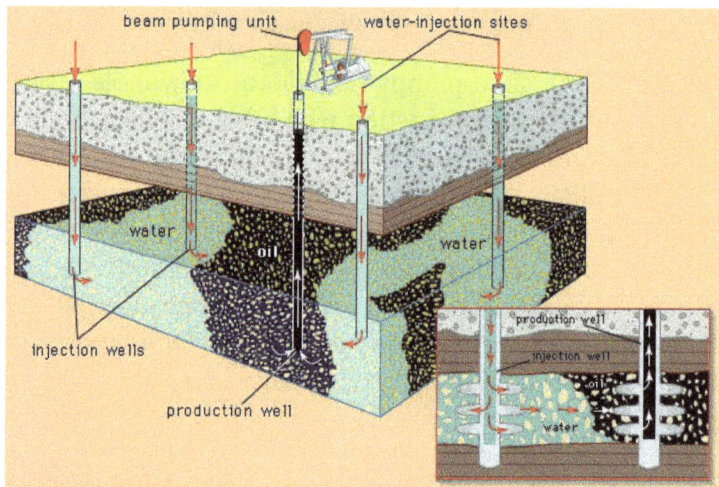

The recovery of petroleum through waterflooding. (Background) Water is pumped into the oil reservoir from several sites around the field; (inset) within the formation, the injected water forces oil toward the production well. Oil and water are pumped to the surface together.

Enhanced Recovery

Enhanced oil recovery (EOR) is designed to accelerate the production of oil from a well. Waterflooding, injecting water to increase the pressure of the reservoir, is one EOR method. Although waterflooding greatly increases recovery from a particular reservoir, it typically leaves up to one-third of the oil in place. Also, shallow reservoirs containing viscous oil do not respond well to waterflooding. Such difficulties have prompted the industry to seek enhanced methods of recovering crude oil supplies. Since many of these methods are directed toward oil that is left behind by water injection, they are often referred to as "tertiary recovery."

Miscible Methods

One method of enhanced recovery is based on the injection of natural gas either at high enough pressure or containing enough petroleum gases in the vapour phase to make the gas and oil miscible. This method leaves little or no oil behind the driving gas, but the relatively low viscosity of the gas can lead to the bypassing of large areas of oil, especially in reservoirs that are not homogeneous. Another enhanced method is intended to recover oil that is left behind by a waterflood by putting a band of soaplike surfactant material ahead of the water. The surfactant creates a very low surface tension between the injected material and the reservoir oil, thus allowing the rock to be "scrubbed" clean. Often, the water behind the surfactant is made viscous by addition of a polymer in order to prevent the water from breaking through and bypassing the surfactant. Surfactant flooding generally works well in noncarbonate rock, but the surfactant material is expensive and large quantities are required. One method that seems to work in carbonate rock is carbon dioxide-enhanced oil recovery (CO_2 EOR), in which carbon dioxide is injected into the rock, either alone or in conjunction with natural gas. CO_2 EOR can greatly improve recovery, but very large quantities of carbon dioxide available at a reasonable price are necessary. Most of the successful projects of this type depend on tapping and transporting (by pipeline) carbon dioxide from underground reservoirs.

In CO_2 EOR, carbon dioxide is injected into an oil-bearing reservoir under high pressure. Oil production relies on the mixtures of gases and the oil, which are strongly dependent on reservoir temperature, pressure, and oil composition. The two main types of CO_2 EOR processes are miscible and immiscible. Miscible CO_2 EOR essentially mixes carbon dioxide with the oil, on which the gas acts as a thinning agent, reducing the oil's viscosity and freeing it from rock pores. The thinned oil is then displaced by another fluid, such as water.

Immiscible CO_2 EOR works on reservoirs with low energy, such as heavy or low-gravity oil reservoirs. Introducing the carbon dioxide into the reservoir creates three mechanisms that work together to energize the reservoir to produce oil: viscosity reduction, oil swelling, and dissolved gas drive, where dissolved gas released from the oil expands to push the oil into the well bore.

CO_2 EOR sources are predominantly taken from naturally occurring carbon dioxide reservoirs. Efforts to use industrial carbon dioxide are advancing in light of potentially detrimental effects of greenhouse gases (such as carbon dioxide) generated by power and chemical plants, for example. However, carbon dioxide capture from combustion processes is costlier than carbon dioxide separation from natural gas reservoirs. Moreover, since plants are rarely located near reservoirs where CO_2 EOR might be useful, the storage and pipeline infrastructure that would be required to deliver the carbon dioxide from plant to reservoir would often be too costly to be feasible.

Thermal Methods

As mentioned above, there are many reservoirs, usually shallow, that contain oil which is too viscous to produce well. Nevertheless, through the application of heat, economical recovery from these reservoirs is possible. Heavy crude oils, which may have a viscosity up to one million times that of water, will show a reduction in viscosity by a factor of 10 for each temperature increase of 50 °C (90 °F). The most successful way to raise the temperature of a reservoir is by the injection of steam. In the most widespread method, called steam cycling, a quantity of steam is injected through a well into a formation and allowed time to condense. Condensation in the reservoir releases the heat of vaporization that was required to create the steam. Then the same well is put into production. After some water production, heated oil flows into the well bore and is lifted to the surface. Often the cycle can be repeated several times in the same well. A less common method involves the injection of steam from one group of wells while oil is continuously produced from other wells.

An alternate method for heating a reservoir involves in situ combustion—the combustion of a part of the reservoir oil in place. Large quantities of compressed air must be injected into the oil zone to support the combustion. The optimal combustion temperature is 500 °C (930 °F). The hot combustion products move through the reservoir to promote oil production. In situ combustion has not seen widespread use.

Gas Cycling

Natural gas reservoirs often contain appreciable quantities of heavier hydrocarbons held in the gaseous state. If reservoir pressure is allowed to decline during gas production, these hydrocarbons will condense in the reservoir to liquefied petroleum gas (LPG) and become unrecoverable. To prevent a decline in pressure, the liquids are removed from the produced gas, and the "dry gas" is put back into the reservoir. This process, called gas cycling, is continued until the optimal quantity of liquids has been recovered. The reservoir pressure is then allowed to decline while the dry gas is produced for sale. In effect, gas cycling defers the use of the natural gas until the liquids have been produced.

Surface Equipment

Water often flows into a well along with oil and natural gas. The well fluids are collected by surface equipment for separation into gas, oil, and water fractions for storage and distribution. The water, which contains salt and other minerals, is usually reinjected into formations that are well separated from freshwater aquifers close to the surface. In many cases it is put back into the formation from which it came. At times, produced water forms an emulsion with the oil or a solid hydrate compound with the gas. In those cases, specially designed treaters are used to separate the three components. The clean crude oil is sent to storage at near atmospheric pressure. Natural gas is usually piped directly to a central gas-processing plant, where "wet gas," or natural gas liquids (NGLs), is removed before it is fed to the consumer pipeline. NGLs are primary feedstock for chemical companies in making various plastics and synthetics. Liquid propane gas (a form of liquefied petroleum gas [LPG]) is a significant component of NGLs and is the source of butane and propane fuels.

Storage and Transport

Offshore production platforms are self-sufficient with respect to power generation and the use of desalinated water for human consumption and operations. In addition, the platforms contain the equipment necessary to process oil prior to its delivery to the shore by pipeline or to a tanker loading facility. Offshore oil production platforms include production separators for separating the produced oil, water, and gas, as well as compressors for any associated gas production. These compressors can also be reused for fuel needs in platform operations, such as water injection pumps, oil and gas export metering, and main oil line pumps. Onshore operations differs from offshore operations in that more space is typically afforded for storage facilities, as well as general access to and from the facilities.

Almost all storage of petroleum is of relatively short duration, lasting only while the oil or gas is awaiting transport or processing. Crude oil, which is stored at or near atmospheric pressure, is usually stored aboveground in cylindrical steel tanks, which may be as large as 30 metres (100 feet) in diameter and 10 metres (33 feet) tall. (Smaller-diameter tanks are used at well sites.) Natural gas and the highly volatile natural gas liquids (NGLs) are stored at higher pressure in steel tanks that are spherical or nearly spherical in shape. Gas is seldom stored, even temporarily, at well sites.

In order to provide supplies when production is lower than demand, longer-term storage of oil and gas is sometimes desirable. This is most often done underground in caverns created inside salt domes or in porous rock formations. Underground reservoirs must be surrounded by nonporous rock so that the oil or gas will stay in place to be recovered later.

Both crude oil and gas must be transported from widely distributed production sites to treatment plants and refineries. Overland movement is largely through pipelines. Crude oil from more isolated wells is collected in tank trucks and taken to pipeline terminals; there is also some transport in specially constructed railroad cars. Pipe used in "gathering lines" to carry oil and gas from wells to a central terminal may be less than 5 cm (2 inches) in diameter. Trunk lines, which carry petroleum over long distances, are as large as 120 cm (48 inches). Where practical, pipelines have been found to be the safest and most economical method to transport petroleum.

Offshore, pipeline infrastructure is often made up of a network of major projects developed by multiple owners. This infrastructure requires a significant initial investment, but its operational life may extend up to 40 years with relatively minor maintenance. The life of the average offshore producing field is 10 years, in comparison, and the pipeline investment is shared so as to manage capacity increases and decreases as new fields are brought online and old ones fade. A stronger justification for sharing ownership is geopolitical risk. Pipelines are often entangled in geopolitical affairs, requiring lengthy planning and advance negotiations designed to appease many interest groups.

The construction of offshore pipelines differs from that of onshore facilities in that the external pressure to the pipe from water requires a greater diameter relative to pipewall thickness. Main onshore transmission lines range from 50 to more than 140 cm (roughly 20 to more than 55 inches) thick. Offshore pipe is limited to diameters of about 91 cm (36 inches) in deep water, though some nearshore pipe is capable of slightly wider diameters; nearshore pipe is as wide as major onshore trunk lines. The range of materials for offshore pipelines is more limited than the range for their onshore counterparts. Seamless pipe and advanced steel alloys are required for offshore operations in order to withstand high pressures and temperatures as depths increase. Basic pipe designs focus on three safety elements: safe installation loads, safe operational loads, and survivability in response to various unplanned conditions, such as sudden changes in undersea topography, severe current changes, and earthquakes.

Although barges are used to transport gathered petroleum from facilities in sheltered inland and coastal waters, overseas transport is conducted in specially designed tanker ships. Tanker capacities vary from less than 100,000 barrels to more than 2,000,000 barrels (4,200,000 to more than 84,000,000 gallons). Tankers that have pressurized and refrigerated compartments also transport compressed liquefied natural gas (LNG) and liquefied petroleum gas (LPG).

Oil tanker: An oil tanker passing through the Kiel Canal in Germany.

Safety and the Environment

Petroleum operations have been high-risk ventures since their inception, and several instances of notable damage to life and property have resulted from oil spills and other petroleum-related accidents as well as acts of sabotage. One of the earliest known incidents was the 1907 Echo Lake fire in downtown Los Angeles, which started when a ruptured oil tank caught fire. Other incidents include the 1978 Amoco Cadiz tanker spill off the coast of Brittany, the opening and ignition of oil wells in 1991 in Iraq and Kuwait during the Persian Gulf War, the 1989 Exxon Valdez spill off the Alaskan coast, and the 2010 Deepwater Horizon oil spill in the Gulf of Mexico. Accidents occur throughout the petroleum production value chain both onshore and offshore. The main causes of these accidents are poor communications, improperly trained workers, failure to enforce safety policies, improper equipment, and rule-based (rather than risk-based) management. These conditions set the stage for oil blowouts (sudden escapes from a well), equipment failures, personal injuries, and deaths of people and wildlife. Preventing accidents requires appreciation and understanding of the risks during each part of petroleum operations.

Human behaviours are the focus for regulatory and legislative health and safety measures. Worker training is designed to cover individual welfare as well as the requirements for processes involving interaction with others—such as lifting and the management of pressure and explosives and other hazardous materials. Licensing is a requirement for many engineers, field equipment operators, and various service providers. For example, offshore crane operators must acquire regulated training and hands-on experience before qualification is granted. However, there are no global standards followed by all countries, states, or provinces. Therefore, it is the responsibility of the operator to seek out and thoroughly understand the local regulations prior to starting operations. The perception that compliance with company standards set within the home country will enable the company to meet all international requirements is incorrect. To facilitate full compliance, employing local staff with detailed knowledge of the local regulations and how they are applied gives confidence to both the visiting company and the enforcing authorities that the operating plans are well prepared.

State-of-the-art operations utilize digital management to remove people from the hazards of surface production processes. This approach, commonly termed "digital oil field (DOF)," essentially allows remote operations by using automated surveillance and control. From a central control room, DOF engineers and operators monitor, evaluate, and respond in advance of issues. This work includes remotely testing or adjusting wells and stopping or starting wells, component valves, fluid separators, pumps, and compressors. Accountability is delegated from the field manager to the process owner, who is typically a leader of a team that is responsible for a specific process, such as drilling, water handling, or well completions. Adopting DOF practices reduces the chances of accidents occurring either on-site or in transit from a well.

Safety during production operations is considered from the bottom of the producing well to the pipeline surface transfer point. Below the surface, wells are controlled by blowout preventers, which the control room or personnel at the well site can use to shut down production when abnormal pressures indicate well integrity or producing zone issues. Remote surveillance using continuous fibre, bottom hole temperature and pressures, and microseismic indicators gives operators early warning signs so that, in most situations, they can take corrective action prior to actuating the blowout preventers. In the case of the 2010 Deepwater Horizon oil spill, the combination of faulty cement installation, mistakes made by managers and crew, and damage to a section of drill pipe that prevented the safety equipment from operating effectively resulted in a blowout that released more than 130 million gallons (about 4.1 million barrels) of oil into the Gulf of Mexico.

Transporting petroleum from the wellhead to the transfer point involves safe handling of the product and monitoring at surface facilities and in the pipeline. Production facilities separate oil, gas, and water and also discard sediments or other undesirable components in preparation for pipeline or tanker transport to the transfer point. Routine maintenance and downtime are scheduled to minimize delays and keep equipment working efficiently. Efficiencies related to rotating equipment performance, for example, are automated to check for declines that may indicate a need for maintenance. Utilization (the ratio of production to total capacity) is checked along with separator and well-test quality to ensure that the range of acceptable performance is met. Sensors attached to pipelines permit remote monitoring and control of pipeline integrity and flow. For example, engineers can remotely regulate the flow of glycol inside pipelines that are building up with hydrates (solid gas crystals formed under low temperatures and pressure). In addition, engineers monitoring sensing equipment can identify potential leaks from corrosion by examining light-scattering data or electric conductivity, and shutdown valves divert flow when leaks are detected. The oldest technique to prevent buildup and corrosion involves using a mechanical device called a "pig," a plastic disk that is run through the pipeline to ream the pipe back to normal operational condition. Another type of pig is the smart pig, which is used to detect problems in the pipeline without shutting down pipeline operations.

With respect to the environment, master operating plans include provisions to minimize waste, including greenhouse gas emissions that may affect climate. Reducing greenhouse gas emissions is part of most operators' plans, which are designed to prevent the emission of flare gas during oil production by sequestering the gas in existing depleted reservoirs and cleaning and reinjecting it into producing reservoirs as an enhanced recovery mechanism. These operations help both the operator and the environment by assisting oil production operations and improving the quality of life for nearby communities.

The final phase in the life of a producing field is abandonment. Wells and producing facilities are scheduled for abandonment only after multiple reviews by management, operations, and engineering departments and by regulatory agencies. Wells are selected for abandonment if their well bores are collapsing or otherwise unsafe. Typically, these wells are plugged with packers that seal off open reservoir zones from their connections with freshwater zones or the surface. In some cases the sections of the wells that span formerly producing zones are cemented but not totally abandoned. This is typical for fields involved in continued production or intended for expansion into new areas. In the case of well abandonment, a workover rig is brought to the field to pull up salvageable materials, such as production tubing, liners, screens, casing, and the wellhead. The workover rig is often a smaller version of a drilling rig, but it is more mobile and constructed without the rotary head. Aside from being involved in the process of well abandonment, workover rigs can be used to reopen producing wells whose downhole systems have failed and pumps or wells that require chemical or mechanical treatments to reinvigorate their producing zones. Upon abandonment, the workover rig is demobilized, all surface connections are removed, and the well site is reconditioned according to its local environment. In most countries, regulatory representatives review and approve abandonments and confirm that the well and the well site are safely closed.

Hydrocarbon Exploration

Hydrocarbon exploration (or oil and gas exploration) is the search by petroleum geologists and geophysicists for deposits of hydrocarbons, particularly petroleum and natural gas, in the Earth using petroleum geology.

Onshore Drilling Rig

Exploration Methods

Visible surface features such as oil seeps, natural gas seeps, pockmarks (underwater craters caused by escaping gas) provide basic evidence of hydrocarbon generation (be it shallow or deep in the Earth). However, most exploration depends on highly sophisticated technology to detect and determine the extent of these deposits using exploration geophysics. Areas thought to contain hydrocarbons are initially subjected to a gravity survey, magnetic survey, passive seismic or regional seismic reflection surveys to detect large-scale features of the sub-surface geology. Features of interest (known as *leads*) are subjected to more detailed seismic surveys which work on the principle of the time it takes for reflected sound waves to travel through matter (rock) of varying densities and using the process of depth conversion to create a profile of the substructure. Finally, when a prospect has been identified and evaluated and passes the oil company's selection criteria, an exploration well is drilled in an attempt to conclusively determine the presence or absence of oil or gas. Offshore the risk can be reduced by using electromagnetic methods.

Oil exploration is an expensive, high-risk operation. Offshore and remote area exploration is generally only undertaken by very large corporations or national governments. Typical shallow shelf oil wells (e.g. North Sea) cost US$10 – 30 million, while deep water wells can cost up to US$100 million plus. Hundreds of smaller companies search for onshore hydrocarbon deposits worldwide, with some wells costing as little as US$100,000.

Elements of a Petroleum Prospect

Mud log in process, a common way to study the rock types when drilling oil wells.

A prospect is a potential trap which geologists believe may contain hydrocarbons. A significant amount of geological, structural and seismic investigation must first be completed to redefine the potential hydrocarbon drill location from a lead to a prospect. Four geological factors have to be present for a prospect to work and if any of them fail neither oil nor gas will be present.

Source Rock

When organic-rich rock such as oil shale or coal is subjected to high pressure and temperature over an extended period of time, hydrocarbons form.

Migration

The hydrocarbons are expelled from source rock by three density-related mechanisms: the newly matured hydrocarbons are less dense than their precursors, which causes over-pressure; the hydrocarbons are lighter, and so migrate upwards due to buoyancy, and the fluids expand as further burial causes increased heating. Most hydrocarbons migrate to the surface as oil seeps, but some will get trapped.

Reservoir

The hydrocarbons are contained in a reservoir rock. This is commonly a porous sandstone or limestone. The oil collects in the pores within the rock although open fractures within non-porous rocks (e.g. fractured granite) may also store hydrocarbons. The reservoir must also be permeable so that the hydrocarbons will flow to surface during production.

Trap

The hydrocarbons are buoyant and have to be trapped within a structural (e.g. Anticline, fault block) or stratigraphic trap. The hydrocarbon trap has to be covered by an impermeable rock known as a seal or cap-rock in order to prevent hydrocarbons escaping to the surface.

Exploration Risk

Hydrocarbon exploration is a high risk investment and risk assessment is paramount for successful project portfolio management. Exploration risk is a difficult concept and is usually defined by assigning confidence to the presence of the imperative geological factors, This confidence is based on data and models and is usually mapped on Common Risk Segment Maps (CRS Maps). High confidence in the presence of imperative geological factors is usually coloured green and low confidence coloured red. Therefore, these maps are also called Traffic Light Maps, while the full procedure is often referred to as Play Fairway Analysis. The aim of such procedures is to force the geologist to objectively assess all different geological factors. Furthermore, it results in simple maps that can be understood by non-geologists and managers to base exploration decisions on.

Terms used in Petroleum Evaluation

- Bright Spot: On a seismic section, coda that have high amplitudes due to a formation containing hydrocarbons.

- Chance of Success: An estimate of the chance of all the elements within a prospect working, described as a probability.

- Dry Hole: A boring that does not contain commercial hydrocarbons.

- Flat Spot: Possibly an oil-water, gas-water or gas-oil contact on a seismic section; flat due to gravity.

- Full Waveform Inversion: A supercomputer technique recently use in conjunction with seismic sensors to explore for petroleum deposits offshore.

- Hydrocarbon in Place: Amount of hydrocarbon likely to be contained in the prospect. This is calculated using the volumetric equation - GRV x N/G x Porosity x Sh/FVF.

- Gross Rock Volume (GRV): Amount of rock in the trap above the hydrocarbon water contact.

- Net Sand: Part of GRV that has the lithological capacity for being a productive zone; i.e. less shale contaminations.

- Net Reserve: Part of net sand that has the minimum reservoir qualities; i.e. minimum porosity and permeability values.

- Net/Gross Ratio (N/G): Proportion of the GRV formed by the reservoir rock (range is 0 to 1).

- Porosity: Percentage of the net reservoir rock occupied by pores (typically 5-35%).

- Hydrocarbon Saturation (Sh): Some of the pore space is filled with water - this must be discounted.

- Formation Volume Factor (FVF): Oil shrinks and gas expands when brought to the surface. The FVF converts volumes at reservoir conditions (high pressure and high temperature) to storage and sale conditions.

- Lead: Potential accumulation is currently poorly defined and requires more data acquisition and evaluation in order to be classified as a prospect.

- Play: An area in which hydrocarbon accumulations or prospects of a given type occur. For example, the shale gas plays in North America include the Barnett, Eagle Ford, Fayetteville, Haynesville, Marcellus, and Woodford, among many others.

- Prospect: A lead which has been more fully evaluated.

- Recoverable hydrocarbons: Amount of hydrocarbon likely to be recovered during production. This is typically 10-50% in an oil field and 50-80% in a gas field.

Licensing

Petroleum resources are typically owned by the government of the host country. In the United States, most onshore (land) oil and gas rights (OGM) are owned by private individuals, in which case oil companies must negotiate terms for a lease of these rights with the individual who owns the OGM. Sometimes this is not the same person who owns the land surface. In most nations the government issues licences to explore, develop and produce its oil and gas resources, which are typically administered by the oil ministry. There are several different types of licence. Oil companies often operate in joint ventures to spread the risk; one of the companies in the partnership is designated the operator who actually supervises the work.

Tax and Royalty

Companies would pay a royalty on any oil produced, together with a profits tax (which can have expenditure offset against it). In some cases there are also various bonuses and ground rents (license fees) payable to the government - for example a signature bonus payable at the start of the licence. Licences are awarded in competitive bid rounds on the basis of either the size of the work programme (number of wells, seismic etc.) or size of the signature bonus.

Production Sharing Contract (PSA)

A PSA is more complex than a Tax/Royalty system - The companies bid on the percentage of the production that the host government receives (this may be variable with the oil price), There is often also participation by the Government owned National Oil Company (NOC). There are also various bonuses to be paid. Development expenditure is offset against production revenue.

Service Contract

This is when an oil company acts as a contractor for the host government, being paid to produce the hydrocarbons.

Reserves and Resources

Resources are hydrocarbons which may or may not be produced in the future. A resource number may be assigned to an undrilled prospect or an unappraised discovery. Appraisal by drilling additional delineation wells or acquiring extra seismic data will confirm the size of the field and lead to project sanction. At this point the relevant government body gives the oil company a production licence which enables the field to be developed. This is also the point at which oil reserves and gas reserves can be formally booked.

Oil and Gas Reserves

Oil and gas reserves are defined as volumes that will be commercially recovered in the future. Reserves are separated into three categories: proved, probable, and possible. To be included in any reserves category, all commercial aspects must have been addressed, which includes government consent. Technical issues alone separate proved from unproved categories. All reserve estimates involve some degree of uncertainty.

Proved reserves are the highest valued category. Proved reserves have a "reasonable certainty" of being recovered, which means a high degree of confidence that the volumes will be recovered. Some industry specialists refer to this as P90, i.e., having a 90% certainty of being produced. The SEC provides a more detailed definition:

Proved oil and gas reserves are those quantities of oil and gas, which, by analysis of geoscience and engineering data, can be estimated with reasonable certainty to be economically producible—from a given date forward, from known reservoirs, and under existing economic conditions, operating methods, and government regulations—prior to the time at which contracts providing the right to operate expire, unless evidence indicates that renewal is reasonably certain, regardless of whether deterministic or probabilistic methods are used for the estimation. The project to extract the hydrocarbons must have commenced or the operator must be reasonably certain that it will commence the project within a reasonable time.

Probable reserves are volumes defined as "less likely to be recovered than proved, but more certain to be recovered than Possible Reserves". Some industry specialists refer to this as P50, i.e., having a 50% certainty of being produced.

Possible reserves are reserves which analysis of geological and engineering data suggests are less likely to be recoverable than probable reserves. Some industry specialists refer to this as P10, i.e., having a 10% certainty of being produced.

The term 1P is frequently used to denote proved reserves; 2P is the sum of proved and probable reserves; and 3P the sum of proved, probable, and possible reserves. The best estimate of recovery from committed projects is generally considered to be the 2P sum of proved and probable reserves. Note that these volumes only refer to currently justified projects or those projects already in development.

Reserve Booking

Oil and gas reserves are the main asset of an oil company. Booking is the process by which they are added to the balance sheet.

In the United States, booking is done according to a set of rules developed by the Society of Petroleum Engineers (SPE). The reserves of any company listed on the New York Stock Exchange have to be stated to the U.S. Securities and Exchange Commission. Reported reserves may be audited by outside geologists, although this is not a legal requirement.

In Russia, companies report their reserves to the State Commission on Mineral Reserves (GKZ).

Hydraulic Fracturing

Hydraulic fracturing (also fracking, fraccing, frac'ing, hydrofracturing or hydrofracking) is a well stimulation technique in which rock is fractured by a pressurized liquid. The process involves the high-pressure injection of 'fracking fluid' (primarily water, containing sand or other proppants suspended with the aid of thickening agents) into a wellbore to create cracks in the deep-rock formations through which natural gas, petroleum, and brine will flow more freely. When the hydraulic pressure is removed from the well, small grains of hydraulic fracturing proppants (either sand or aluminium oxide) hold the fractures open.

Hydraulic fracturing began as an experiment in 1947, and the first commercially successful application followed in 1950. As of 2012, 2.5 million "frac jobs" had been performed worldwide on oil and gas wells; over one million of those within the U.S. Such treatment is generally necessary to achieve adequate flow rates in shale gas, tight gas, tight oil, and coal seam gas wells. Some hydraulic fractures can form naturally in certain veins or dikes.

Hydraulic fracturing is highly controversial in many countries. Its proponents advocate the economic benefits of more extensively accessible hydrocarbons, as well as replacing coal with natural gas, which burns cleaner and emits half as much carbon dioxide (CO_2). Opponents of fracking argue that these are outweighed by the potential environmental impacts, which include risks of ground water and surface water contamination, noise and air pollution, and the triggering of earthquakes, along with the resulting hazards to public health and the environment.

Methane leakage is also a problem directly associated with hydraulic fracturing, as a Environmental Defense Fund (EDF) report in the US highlights, where the leakage rate in Pennsylvania during extensive testing and analysis was found to be approximately 10%, or over five times the reported figures. This leakage rate is considered representative of the hydraulic fracturing industry in the US generally. The EDF have recently announced a satellite mission to further locate and measure methane emissions.

Increases in seismic activity following hydraulic fracturing along dormant or previously unknown faults are sometimes caused by the deep-injection disposal of hydraulic fracturing flowback (a byproduct of hydraulically fractured wells), and produced formation brine (a byproduct of both fractured and nonfractured oil and gas wells). For these reasons, hydraulic fracturing is under international scrutiny, restricted in some countries, and banned altogether in others. The European Union is drafting regulations that would permit the controlled application of hydraulic fracturing.

Geology

Halliburton fracturing operation in the Bakken Formation, North Dakota, United States.

A fracturing operation in progress.

Mechanics

Fracturing rocks at great depth frequently becomes suppressed by pressure due to the weight of the overlying rock strata and the cementation of the formation. This suppression process is particularly significant in "tensile" (Mode 1) fractures which require the walls of the fracture to move against this pressure. Fracturing occurs when effective stress is overcome by the pressure of fluids within the rock. The minimum principal stress becomes tensile and exceeds the tensile strength of the material. Fractures formed in this way are generally oriented in a plane perpendicular to the minimum principal stress, and for this reason, hydraulic fractures in well bores can be used to determine the orientation of stresses. In natural examples, such as dikes or vein-filled fractures, the orientations can be used to infer past states of stress.

Veins

Most mineral vein systems are a result of repeated natural fracturing during periods of relatively high pore fluid pressure. The impact of high pore fluid pressure on the formation process of mineral vein systems is particularly evident in "crack-seal" veins, where the vein material is part of a series of discrete fracturing events, and extra vein material is deposited on each occasion. One example of long-term repeated natural fracturing is in the effects of seismic activity. Stress levels rise and fall episodically, and earthquakes can cause large volumes of connate water to be expelled from fluid-filled fractures. This process is referred to as "seismic pumping".

Dikes

Minor intrusions in the upper part of the crust, such as dikes, propagate in the form of fluid-filled cracks. In such cases, the fluid is magma. In sedimentary rocks with a significant water content, fluid at fracture tip will be steam.

Process

Hydraulic fracturing is a process to stimulate a natural gas, oil, or geothermal well to maximize extraction. The EPA defines the broader process to include acquisition of source water, well construction, well stimulation, and waste disposal.

Method

A hydraulic fracture is formed by pumping fracturing fluid into a wellbore at a rate sufficient to increase pressure at the target depth (determined by the location of the well casing perforations), to exceed that of the fracture *gradient* (pressure gradient) of the rock. The fracture gradient is defined as pressure increase per unit of depth relative to density, and is usually measured in pounds per square inch, per square foot, or bars. The rock cracks, and the fracture fluid permeates the rock extending the crack further, and further, and so on. Fractures are localized as pressure drops off with the rate of frictional loss, which is relative to the distance from the well. Operators typically try to maintain "fracture width", or slow its decline following treatment, by introducing a proppant into the injected fluid – a material such as grains of sand, ceramic, or other particulate, thus preventing the fractures from closing when injection is stopped and pressure removed. Consideration of proppant strength and prevention of proppant failure becomes more important at greater depths where pressure and stresses on fractures are higher. The propped fracture is permeable enough to allow the flow of gas, oil, salt water and hydraulic fracturing fluids to the well.

During the process, fracturing fluid leakoff (loss of fracturing fluid from the fracture channel into the surrounding permeable rock) occurs. If not controlled, it can exceed 70% of the injected volume. This may result in formation matrix damage, adverse formation fluid interaction, and altered fracture geometry, thereby decreasing efficiency.

The location of one or more fractures along the length of the borehole is strictly controlled by various methods that create or seal holes in the side of the wellbore. Hydraulic fracturing is performed in cased wellbores, and the zones to be fractured are accessed by perforating the casing at those locations.

Hydraulic-fracturing equipment used in oil and natural gas fields usually consists of a slurry blender, one or more high-pressure, high-volume fracturing pumps (typically powerful triplex or quintuplex pumps) and a monitoring unit. Associated equipment includes fracturing tanks, one or more units for storage and handling of proppant, high-pressure treating iron, a chemical additive unit (used to accurately monitor chemical addition), low-pressure flexible hoses, and many gauges and meters for flow rate, fluid density, and treating pressure. Chemical additives are typically 0.5% of the total fluid volume. Fracturing equipment operates over a range of pressures and injection rates, and can reach up to 100 megapascals (15,000 psi) and 265 litres per second (9.4 cu ft/s) (100 barrels per minute).

Well Types

A distinction can be made between conventional, low-volume hydraulic fracturing, used to stimulate high-permeability reservoirs for a single well, and unconventional, high-volume hydraulic fracturing, used in the completion of tight gas and shale gas wells. High-volume hydraulic fracturing usually requires higher pressures than low-volume fracturing; the higher pressures are needed to push out larger volumes of fluid and proppant that extend farther from the borehole.

Horizontal drilling involves wellbores with a terminal drillhole completed as a "lateral" that extends parallel with the rock layer containing the substance to be extracted. For example, laterals extend 1,500 to 5,000 feet (460 to 1,520 m) in the Barnett Shale basin in Texas, and up to 10,000 feet (3,000 m) in the Bakken formation in North Dakota. In contrast, a vertical well only accesses the thickness of the rock layer, typically 50–300 feet (15–91 m). Horizontal drilling reduces surface disruptions as fewer wells are required to access the same volume of rock.

Drilling often plugs up the pore spaces at the wellbore wall, reducing permeability at and near the wellbore. This reduces flow into the borehole from the surrounding rock formation, and partially seals off the borehole from the surrounding rock. Low-volume hydraulic fracturing can be used to restore permeability.

Fracturing Fluids

Water tanks preparing for hydraulic fracturing.

The main purposes of fracturing fluid are to extend fractures, add lubrication, change gel strength, and to carry proppant into the formation. There are two methods of transporting proppant in the fluid – high-rate and high-viscosity. High-viscosity fracturing tends to cause large dominant fractures, while high-rate (slickwater) fracturing causes small spread-out micro-fractures.

Water-soluble gelling agents (such as guar gum) increase viscosity and efficiently deliver proppant into the formation.

Example of high pressure manifold combining pump flows before injection into well.

Fluid is typically a slurry of water, proppant, and chemical additives. Additionally, gels, foams, and compressed gases, including nitrogen, carbon dioxide and air can be injected. Typically, 90% of the fluid is water and 9.5% is sand with chemical additives accounting to about 0.5%. However, fracturing fluids have been developed using liquefied petroleum gas (LPG) and propane in which water is unnecessary.

The proppant is a granular material that prevents the created fractures from closing after the fracturing treatment. Types of proppant include silica sand, resin-coated sand, bauxite, and man-made ceramics. The choice of proppant depends on the type of permeability or grain strength needed. In some formations, where the pressure is great enough to crush grains of natural silica sand, higher-strength proppants such as bauxite or ceramics may be used. The most commonly used proppant is silica sand, though proppants of uniform size and shape, such as a ceramic proppant, are believed to be more effective.

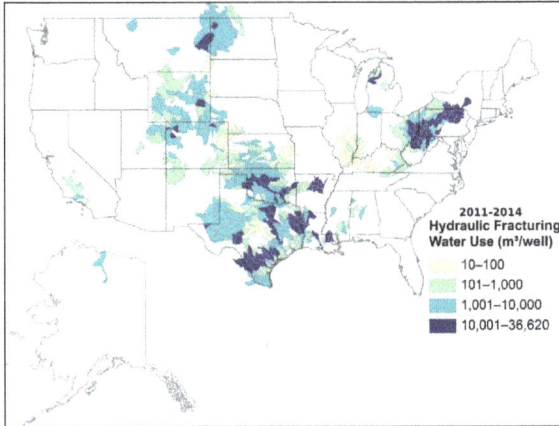

USGS map of water use from hydraulic fracturing between 2011 and 2014.
One cubic meter of water is 264.172 gallons.

The fracturing fluid varies depending on fracturing type desired, and the conditions of specific wells being fractured, and water characteristics. The fluid can be gel, foam, or slickwater-based. Fluid choices are tradeoffs: more viscous fluids, such as gels, are better at keeping proppant in suspension; while less-viscous and lower-friction fluids, such as slickwater, allow fluid to be pumped at higher rates, to create fractures farther out from the wellbore. Important material properties of the fluid include viscosity, pH, various rheological factors, and others.

Water is mixed with sand and chemicals to create hydraulic facturing fluid. Approximately 40,000 gallons of chemicals are used per fracturing. A typical fracture treatment uses between 3 and 12 additive chemicals. Although there may be unconventional fracturing fluids, typical chemical additives can include one or more of the following:

- Acids—Hydrochloric acid or acetic acid is used in the pre-fracturing stage for cleaning the perforations and initiating fissure in the near-wellbore rock.

- Sodium chloride (salt)—Delays breakdown of gel polymer chains.

- Polyacrylamide and other friction reducers decrease turbulence in fluid flow and pipe friction, thus allowing the pumps to pump at a higher rate without having greater pressure on the surface.

- Ethylene glycol—Prevents formation of scale deposits in the pipe.

- Borate salts—Used for maintaining fluid viscosity during the temperature increase.

- Sodium and potassium carbonates—Used for maintaining effectiveness of crosslinkers.

- Anaerobic, Biocide, BIO—Glutaraldehyde used as disinfectant of the water (bacteria elimination).

- Guar gum and other water-soluble gelling agents—Increases viscosity of the fracturing fluid to deliver proppant into the formation more efficiently.

- Citric acid—Used for corrosion prevention.

- Isopropanol—Used to winterize the chemicals to ensure it doesn't freeze.

The most common chemical used for hydraulic fracturing in the United States in 2005–2009 was methanol, while some other most widely used chemicals were isopropyl alcohol, 2-butoxyethanol, and ethylene glycol.

Typical fluid types are:

- Conventional linear gels. These gels are cellulose derivative (carboxymethyl cellulose, hydroxyethyl cellulose, carboxymethyl hydroxyethyl cellulose, hydroxypropyl cellulose, hydroxyethyl methyl cellulose), guar or its derivatives (hydroxypropyl guar, carboxymethyl hydroxypropyl guar), mixed with other chemicals.

- Borate-crosslinked fluids. These are guar-based fluids cross-linked with boron ions (from aqueous borax/boric acid solution). These gels have higher viscosity at pH 9 onwards and are used to carry proppant. After the fracturing job, the pH is reduced to 3–4 so that the cross-links are broken, and the gel is less viscous and can be pumped out.

- Organometallic-crosslinked fluids – zirconium, chromium, antimony, titanium salts – are known to crosslink guar-based gels. The crosslinking mechanism is not reversible, so once the proppant is pumped down along with cross-linked gel, the fracturing part is done. The gels are broken down with appropriate breakers.

- Aluminium phosphate-ester oil gels. Aluminium phosphate and ester oils are slurried to form cross-linked gel. These are one of the first known gelling systems.

For slickwater fluids the use of sweeps is common. Sweeps are temporary reductions in the proppant concentration, which help ensure that the well is not overwhelmed with proppant. As the fracturing process proceeds, viscosity-reducing agents such as oxidizers and enzyme breakers are sometimes added to the fracturing fluid to deactivate the gelling agents and encourage flowback.

Such oxidizers react with and break down the gel, reducing the fluid's viscosity and ensuring that no proppant is pulled from the formation. An enzyme acts as a catalyst for breaking down the gel. Sometimes pH modifiers are used to break down the crosslink at the end of a hydraulic fracturing job, since many require a pH buffer system to stay viscous. At the end of the job, the well is commonly flushed with water under pressure (sometimes blended with a friction reducing chemical.) Some (but not all) injected fluid is recovered. This fluid is managed by several methods, including underground injection control, treatment, discharge, recycling, and temporary storage in pits or containers. New technology is continually developing to better handle waste water and improve re-usability.

Fracture Monitoring

Measurements of the pressure and rate during the growth of a hydraulic fracture, with knowledge of fluid properties and proppant being injected into the well, provides the most common and simplest method of monitoring a hydraulic fracture treatment. This data along with knowledge of the underground geology can be used to model information such as length, width and conductivity of a propped fracture.

Injection of radioactive tracers along with the fracturing fluid is sometimes used to determine the injection profile and location of created fractures. Radiotracers are selected to have the readily detectable radiation, appropriate chemical properties, and a half life and toxicity level that will minimize initial and residual contamination. Radioactive isotopes chemically bonded to glass (sand) and resin beads may also be injected to track fractures. For example, plastic pellets coated with 10 GBq of Ag-110mm may be added to the proppant, or sand may be labelled with Ir-192, so that the proppant's progress can be monitored. Radiotracers such as Tc-99m and I-131 are also used to measure flow rates. The Nuclear Regulatory Commission publishes guidelines which list a wide range of radioactive materials in solid, liquid and gaseous forms that may be used as tracers and limit the amount that may be used per injection and per well of each radionuclide.

A new technique in well-monitoring involves fiber-optic cables outside the casing. Using the fiber optics, temperatures can be measured every foot along the well – even while the wells are being fracked and pumped. By monitoring the temperature of the well, engineers can determine how much hydraulic fracturing fluid different parts of the well use as well as how much natural gas or oil they collect, during hydraulic fracturing operation and when the well is producing.

Microseismic Monitoring

For more advanced applications, microseismic monitoring is sometimes used to estimate the size and orientation of induced fractures. Microseismic activity is measured by placing an array of geophones in a nearby wellbore. By mapping the location of any small seismic events associated with the growing fracture, the approximate geometry of the fracture is inferred. Tiltmeter arrays deployed on the surface or down a well provide another technology for monitoring strain.

Microseismic mapping is very similar geophysically to seismology. In earthquake seismology, seismometers scattered on or near the surface of the earth record S-waves and P-waves that are released during an earthquake event. This allows for motion along the fault plane to be estimated and its location in the Earth's subsurface mapped. Hydraulic fracturing, an increase in formation stress proportional to the net fracturing pressure, as well as an increase in pore pressure due to leakoff. Tensile stresses are generated ahead of the fracture's tip, generating large amounts of shear stress. The increases in pore water pressure and in formation stress combine and affect weaknesses near the hydraulic fracture, like natural fractures, joints, and bedding planes.

Different methods have different location errors and advantages. Accuracy of microseismic event mapping is dependent on the signal-to-noise ratio and the distribution of sensors. Accuracy of events located by seismic inversion is improved by sensors placed in multiple azimuths from the monitored borehole. In a downhole array location, accuracy of events is improved by being close to the monitored borehole (high signal-to-noise ratio).

Monitoring of microseismic events induced by reservoir stimulation has become a key aspect in evaluation of hydraulic fractures, and their optimization. The main goal of hydraulic fracture monitoring is to completely characterize the induced fracture structure, and distribution of conductivity within a formation. Geomechanical analysis, such as understanding a formations material properties, in-situ conditions, and geometries, helps monitoring by providing a better definition of the environment in which the fracture network propagates. The next task is to know the location of proppant within the fracture and the distribution of fracture conductivity. This can be monitored using multiple types of techniques to finally develop a reservoir model than accurately predicts well performance.

Horizontal Completions

Since the early 2000s, advances in drilling and completion technology have made horizontal wellbores much more economical. Horizontal wellbores allow far greater exposure to a formation than conventional vertical wellbores. This is particularly useful in shale formations which do not have sufficient permeability to produce economically with a vertical well. Such wells, when drilled onshore, are now usually hydraulically fractured in a number of stages, especially in North America. The type of wellbore completion is used to determine how many times a formation is fractured, and at what locations along the horizontal section.

In North America, shale reservoirs such as the Bakken, Barnett, Montney, Haynesville, Marcellus, and most recently the Eagle Ford, Niobrara and Utica shales are drilled horizontally through the producing interval(s), completed and fractured. The method by which the fractures are placed along the wellbore is most commonly achieved by one of two methods, known as "plug and perf" and "sliding sleeve".

The wellbore for a plug-and-perf job is generally composed of standard steel casing, cemented or uncemented, set in the drilled hole. Once the drilling rig has been removed, a wireline truck is used to perforate near the bottom of the well, and then fracturing fluid is pumped. Then the wireline truck sets a plug in the well to temporarily seal off that section so the next section of the wellbore can be treated. Another stage is pumped, and the process is repeated along the horizontal length of the wellbore.

The wellbore for the sliding sleeve technique is different in that the sliding sleeves are included at set spacings in the steel casing at the time it is set in place. The sliding sleeves are usually all closed at this time. When the well is due to be fractured, the bottom sliding sleeve is opened using one of several activation techniques and the first stage gets pumped. Once finished, the next sleeve is opened, concurrently isolating the previous stage, and the process repeats. For the sliding sleeve method, wireline is usually not required.

Sleeves

These completion techniques may allow for more than 30 stages to be pumped into the horizontal section of a single well if required, which is far more than would typically be pumped into a vertical well that had far fewer feet of producing zone exposed.

Uses

Hydraulic fracturing is used to increase the rate at which fluids, such as petroleum, water, or natural gas can be recovered from subterranean natural reservoirs. Reservoirs are typically porous sandstones, limestones or dolomite rocks, but also include "unconventional reservoirs" such as shale rock or coal beds. Hydraulic fracturing enables the extraction of natural gas and oil from rock formations deep below the earth's surface (generally 2,000–6,000 m (5,000–20,000 ft)), which is greatly below typical groundwater reservoir levels. At such depth, there may be insufficient permeability or reservoir pressure to allow natural gas and oil to flow from the rock into the wellbore at high economic return. Thus, creating conductive fractures in the rock is instrumental in extraction from naturally impermeable shale reservoirs. Permeability is measured in the microdarcy to nanodarcy range. Fractures are a conductive path connecting a larger volume of reservoir to the well. So-called "super fracking," creates cracks deeper in the rock formation to release more oil and gas, and increases efficiency. The yield for typical shale bores generally falls off after the first year or two, but the peak producing life of a well can be extended to several decades.

While the main industrial use of hydraulic fracturing is in stimulating production from oil and gas wells, hydraulic fracturing is also applied:

- To stimulate groundwater wells.
- To precondition or induce rock cave-ins mining.
- As a means of enhancing waste remediation, usually hydrocarbon waste or spills.
- To dispose waste by injection deep into rock.
- To measure stress in the Earth.
- For electricity generation in enhanced geothermal systems.
- To increase injection rates for geologic sequestration of CO_2.

Since the late 1970s, hydraulic fracturing has been used, in some cases, to increase the yield of drinking water from wells in a number of countries, including the United States, Australia, and South Africa.

Economic Effects

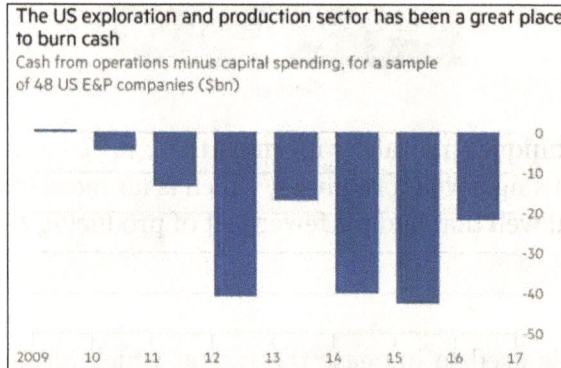

The US exploration and production sector has been a great place to burn cash

Cash from operations minus capital spending, for a sample of 48 US E&P companies ($bn)

Production costs for unconventional oil and gas continue to outweigh profits.

Hydraulic fracturing has been seen as one of the key methods of extracting unconventional oil and unconventional gas resources. According to the International Energy Agency, the remaining technically recoverable resources of shale gas are estimated to amount to 208 trillion cubic metres (7,300 trillion cubic feet), tight gas to 76 trillion cubic metres (2,700 trillion cubic feet), and coalbed methane to 47 trillion cubic metres (1,700 trillion cubic feet). As a rule, formations of these resources have lower permeability than conventional gas formations. Therefore, depending on the geological characteristics of the formation, specific technologies such as hydraulic fracturing are required. Although there are also other methods to extract these resources, such as conventional drilling or horizontal drilling, hydraulic fracturing is one of the key methods making their extraction economically viable. The multi-stage fracturing technique has facilitated the development of shale gas and light tight oil production in the United States and is believed to do so in the other countries with unconventional hydrocarbon resources.

A large majority of studies indicate that hydraulic fracturing in the United States has had a strong positive economic benefit so far. The Brookings Institution estimates that the benefits of Shale Gas alone has led to a net economic benefit of $48 billion per year. Most of this benefit is within the consumer and industrial sectors due to the significantly reduced prices for natural gas. Other studies have suggested that the economic benefits are outweighed by the externalities and that the levelized cost of electricity (LCOE) from less carbon and water intensive sources is lower.

The primary benefit of hydraulic fracturing is to offset imports of natural gas and oil, where the cost paid to producers otherwise exits the domestic economy. However, shale oil and gas is highly subsidisied in the US, and has not yet covered production costs - meaning that the cost of hydraulic fracturing is paid for in income taxes, and in many cases is up to double the cost paid at the pump.

Research suggests that hydraulic fracturing wells have an adverse impact on agricultural productivity in the vicinity of the wells. One paper found "that productivity of an irrigated crop decreases by 5.7% when a well is drilled during the agriculturally active months within 11–20 km radius of a producing township. This effect becomes smaller and weaker as the distance between township and wells increases." The findings imply that the introduction of hydraulic fracturing wells to Alberta cost the province $14.8 million in 2014 due to the decline in the crop productivity. The Energy Information Administration of the US Department of Energy estimates that 45% of US gas supply will come from shale gas by 2035 (with the vast majority of this replacing conventional gas, which has a lower greenhouse-gas footprint).

Oil and Gas Exploration Methods

Exploration is a critical upstream activity in the oil and gas industry. Due to increasing demands from developed and developing countries, exploring oil or gas depositions remains an important and ongoing global pursuit. However, exploration can only commence once costs and benefits have been assessed and once the proper type of oil and gas agreement or contract has been determined.

Nonetheless, there are three general classifications or types of methods and strategies used in oil and gas exploration.These are surface methods, gravity and magnetic surveys, and seismographic methods. Note that these methods are specifically used to determine the presence of an oil or gas under a particular land area or a specified area within the seabed.

Surface Methods

Oil and gas exploration using surface methods are based on either one of the two principles. The first is to survey the geological feature of the surface to determine sedimentary rock formations and repeated folds and faults or salt domes in subsurface rock formation. Note that the rocks that contain oil and gas are all sedimentary rocks.

The second is to determine hydrocarbon seepage on the surface of the earth. Oil seeps in low areas, thus becoming tar-like deposits that can be visually observed and further confirmed using lab analysis. Large deposits of oil underground do not leak, however. On the other hand, gas is not visible but its concentration in the air or seawater can be detected using special instruments.

Gravity and Magnetic Surveys

Small differences in gravitational force can be picked by an instrument known as gravimeter and can serve further as a basis for assessing the presence of oil or gas deposits. Remember that gravitational force is slightly greater on surfaces close to dense rock formation and it is slightly weaker on surfaces with slat domes underneath.

Using an instrument called magnetometers that can also be attached to an aircraft or sea vessel, properties of rock and rock formation or anomalies can be assessed and mapped based on the response to magnetic field measurements and electrical resistivity.

Seismographic Methods

Both surface methods and gravity or magnetic surveys can map out large geologic anomalies such as faults and folds, salt domes, and anticlines. However, the problem with these oil and gas exploration methods is that they cannot produce detailed images of smaller and inconspicuous areas.

Seismographic methods use sound waves to produce detailed images of underground rock formations. These methods can involve invasive and destructive techniques however, especially when drilling shallow holes on the surface to place microphones or listening devices to reflect and transmit sound signals.

The aforementioned methods and strategies do not guarantee the presence of oil or gas deposits. Results from using either one or more of these methods merely indicate the presence or absence of favorable underground formations or geographic features and characteristics that are conducive for the accumulation of hydrocarbons. It is also important to stress the difference between resources and reserves, especially in classifying the viability of the suspected or proven oil and gas deposits in a particular area.

Spontaneous Potential Logging

The SP curve is a continuous recording vs. depth of the electrical potential difference between a movable electrode in the borehole and a surface electrode. Adjacent to shales, SP readings usually define a straight line known as the shale baseline. Next to permeable formations, the curve departs from the shale baseline; in thick permeable beds, these excursions reach a constant departure from the shale baseline, defining the "sand line." The deflection may be either to the left (negative) or to the right (positive), depending on the relative salinities of the formation water and the mud filtrate. If the formation-water salinity is greater than the mud-filtrate salinity (the more common case), the deflection is to the left.

The relevant features of the SP curve are its shape and the size of its departure from the shale baseline. Because the absolute reading and position of the shale baseline on the log are irrelevant, the SP sensitivity scale and shale-baseline position are selected by the logging engineer for convenience. The SP log is typically scaled at 100 mV per log track. If the resistivities of the mud filtrate and formation water are similar, the SP deflections are small and the curve is rather featureless. An SP curve cannot be recorded in holes filled with nonconductive muds, such as oil-based muds (OBMs).

Deflections of the SP curve are the result of electrochemical and electrokinetic potentials in the formations that cause electric currents to flow in the mud in the borehole.

Electrochemical Component

Membrane Potential

The structure of clay minerals in shales and the concentration of negative electric charges on the clay particle surfaces give shales a selective permeability to electrically charged ions. Most shales act as "cationic membranes" that are permeable to positively charged ions (cations) and impermeable to negative ions (anions).

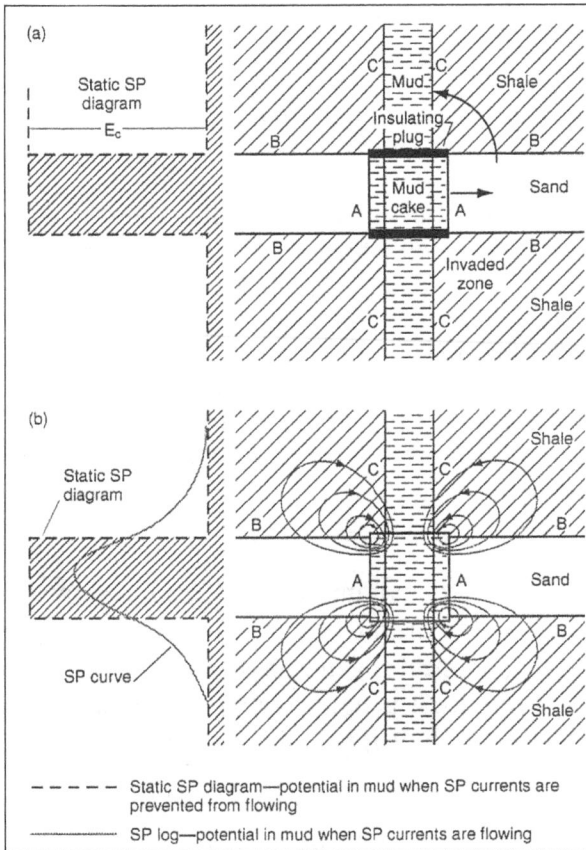

Schematic representation of potential and current distribution in and around a permeable bed.

The upper part of figure shows saline formation water in a sandstone formation and mud in the borehole separated by a shale. Sodium chloride, which is usually present in both the formation water and the drilling mud, separates into charged ions (Na^+ and Cl^-) in solution in water. The Na^+ and Cl^- ions tend to migrate from a more-concentrated to a less-concentrated solution, but because the intervening shale is a cationic membrane, impervious to Cl^- ions, only the Na^+ ions can migrate. If, as usual, the formation water is a more concentrated NaCl solution than the mud, there is a net flow of positive ions through the shale from the sandstone to the borehole. This corresponds to a positive electric current in the same direction (indicated by the curved arrow) driven by an electric potential, or electromotive force (EMF), across the shale. Because the shale acts as an ion-selective membrane, the electric potential is known as the membrane potential.

Liquid-junction Potential

At the edge of the invaded zone, where the mud filtrate and formation water are in direct contact, Na^+ and Cl^- ions can move freely from one solution to the other. But Cl^- ions are smaller and have greater mobility than Na^+ ions, so the net diffusion of ions from the more-concentrated formation water to the less-concentrated mud filtrate includes a greater number of Cl^- ions than Na^+ ions. This is equivalent to a positive current flow in the opposite direction (indicated by the straight arrow at A in figure).

The current flowing across the junction between solutions of different salinity is driven by an EMF called the liquid-junction potential. The magnitude of the liquid-junction potential is only approximately one-fifth of the membrane potential.

If the solutions contain substantial amounts of salts other than NaCl, the value of K at 77 °F may not be 71. If the permeable formation contains some shale or dispersed clay, the total electrochemical potential, and therefore the SP deflections, is reduced.

Electrokinetic Component

An electrokinetic potential (E_k, also called the streaming potential or electrofiltration potential) is produced when an electrolyte flows through a permeable medium. The size of the electrokinetic potential is determined mainly by the differential pressure producing the flow and the resistivity of the electrolyte.

In the borehole, the electrokinetic potential E_{kmc} is produced by the flow of mud filtrate through the mudcake deposited on the borehole wall opposite permeable formations. Little or no electrokinetic potential is generated across the permeable formation itself because the differential pressure is usually low. The electrokinetic potential Eksh may, however, be produced across a shale if it has any permeability.

Typically, E_{kmc} and E_{ksh} are similar in magnitude, and the net electrokinetic contribution to the SP deflection is negligible. If the formation water is fairly saline (resistivity less than 0.1 ohm·m) and the differential pressure is in the normal range of only a few hundred psi, the contribution of the electrokinetic potential can usually be ignored.

Electrokinetic effects may be significant in highly depleted formations or when heavy drilling muds are used because of unusually large differential pressures. Significant electrokinetic effects may also occur in very-low-permeability formations, where an appreciable part of the pressure differential occurs in the formation itself, especially if little or no mudcake is formed. If the formation water is brackish, the mud is resistive, and the low-permeability formation is clean and has some porosity, the electrokinetic effect could be as large as −200 mV.

SP and Permeability

The movement of ions, essential to develop an SP, is possible only in formations with some permeability, however small—a small fraction of a millidarcy is sufficient. There is no direct relationship between the magnitude of the SP deflection and the value of either the formation's permeability or its porosity.

Static SP

The lower part of figure shows SP currents in the borehole and formations. The current directions indicated correspond to the more usual case of formation-water salinity greater than mud-filtrate salinity, producing a potential by the permeable bed lower than the potential by the shale. This corresponds to a deflection to the left on the SP log by the permeable bed.

If the mud-filtrate salinity is greater than the formation-water salinity, the currents flow in the opposite direction, producing positive SP deflections. If the salinities of the mud filtrate and formation water are similar, no SP is generated.

The SP currents flow through four different media:

- Borehole fluid.

- The invaded zone.

- The uninvaded part of the permeable formation.

- Surrounding shales.

The SP log measures only the potential drop from the SP currents in the borehole fluid, which may not represent the total SP because there are also potential drops in the formation. If the currents could be interrupted by hypothetical insulating plugs, the potential observed in the mud would be the total spontaneous potential. This idealized SP deflection is called the static SP (or SSP). The SP deflection practically reaches the SSP in a thick, clean formation.

The borehole presents a much smaller cross-sectional area to current flow than the formations around it, so the resistance of the borehole part of the SP current loop is much higher than the formation part. Nearly all the SP potential drop, therefore, occurs in the borehole if formation resistivities are low-to-moderate and formation beds are thick, so, in practice, the recorded SP deflection approaches the static SP value in thick, permeable beds.

Determination of SSP

To determine the SSP, a sand line is drawn through the maximum (usually negative) excursions of the SP curve adjacent to the thickest permeable beds. A shale baseline is drawn through the SP through the intervening shale beds. The separation of the sand line from the shale baseline, measured in mV, is the SSP. Any SP anomalies are discounted.

If there are no thick, clean, permeable invaded beds in the zone under study, the SP reading can be corrected for the effects of bed thickness and invasion to estimate the SSP by using charts available from service companies.

Shape of the SP Curve

The slope of the SP curve is proportional to the intensity of the SP currents in the borehole at that depth. Because the current intensity is highest at the boundaries of the permeable formation, the slope of the SP curve is at a maximum, and an inflection point occurs at these bed boundaries.

The shape of the SP curve and the amplitude of its deflection in permeable beds depend on the following factors: thickness and true formation resistivity of the permeable bed, resistivity of the flushed zone (R_{xo}) and diameter d_i, resistivity of the adjacent shale bed (R_s), and resistivity of the mud and the diameter of the borehole (d_h).

Figure shows examples of SP curves computed for $R_t = R_s = R_m$ (on the left) and $R_t = R_s = 21R_m$ (in the center). In the first case $(R_t = R_s = R_m)$, the SP curve gives a much sharper definition of the boundaries of the permeable beds, and the SP deflections approach the SSP value more closely than in the case where the formation-to-mud resistivity ratio is 21.

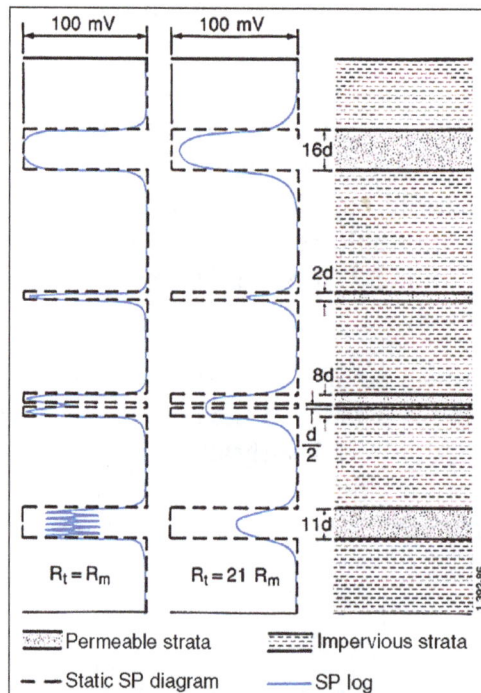

SP curve in beds of different thickness for $R_t = R_m$ (left) and $R_t = 21 R_m$ (center).

SP Anomalies

The SP curve may be difficult to interpret and use for Rw determination because it does not always behave ideally. The following are a few cases of apparently anomalous SP responses.

Highly Resistive Formations

Highly resistive formations between some shales and permeable beds can significantly alter the distribution of the SP currents and change the expected shape of the SP curve. The currents shown flowing from shale bed Sh 1 toward permeable bed P2 in figure are largely confined to the borehole by the high resistivity of the formation separating Sh1 and P2. The current in the borehole over this interval is constant, so for a constant borehole diameter, the SP curve is a straight line inclined to the shale baseline.

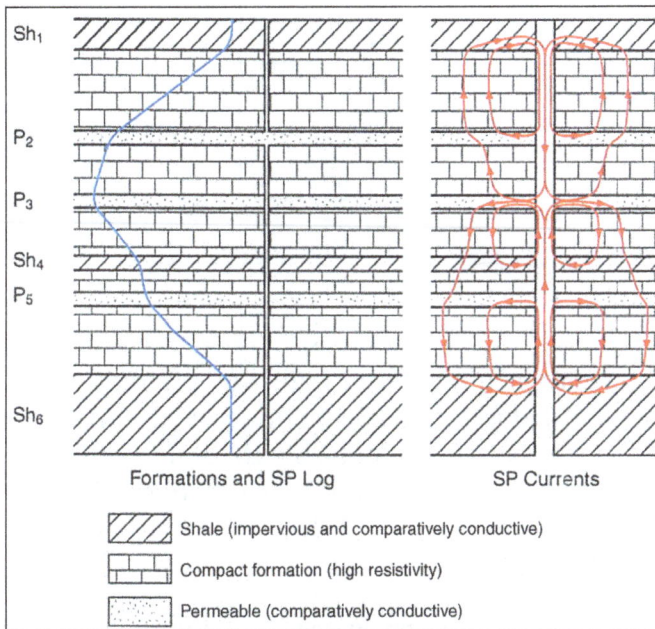

Schematic representations of SP phenomena in highly resistive formations.

The SP curve consists of straight portions adjacent to the high-resistivity zones with a change of slope at each more conductive permeable interval (the SP curve is concave toward the shale line) and opposite every shale bed (the SP curve is convex toward the shale line). Defining permeable bed boundaries from the SP log is difficult in the vicinity of highly resistive formations.

Shale-baseline Shifts

A shift of the shale baseline can occur when formation waters of different salinities are separated by a shale bed that is not a perfect cationic membrane. Figure shows an SP log recorded in a series of sandstones (B, D, F, and H) separated by thin shales or shaly sandstones (C, E, and G). The SSP of Sandstone B is −42 mV. Shale C is not a perfect cationic membrane, and the SP curve does not return to the shale baseline defined by Shale A. A new shale baseline defined by Shale E gives SP deflections of +44 mV in Sandstone D and −23 mV in Sandstone F.

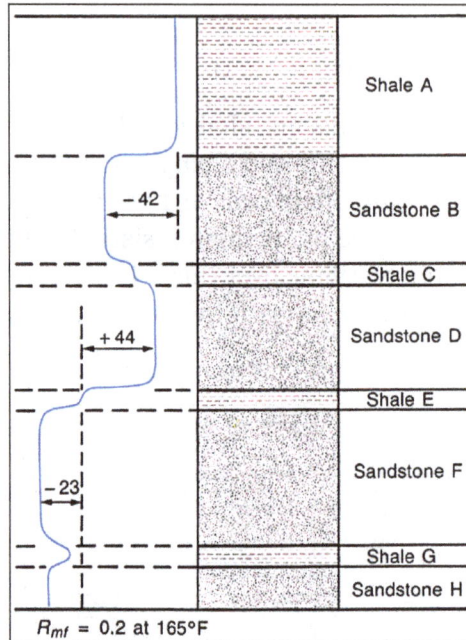

SP baseline shift.

Baseline shifts also occur when formation waters of different salinities are separated by an impermeable layer that is not a shale. In this case, the SP curve shows little or no variation at the level of the change in salinity, but the deflections at the upper and lower shale boundaries are different and may even have different polarities.

Invasion-related Anomalies

If the mud filtrate and the formation water have significantly different salinities, and therefore different densities, gravity-induced fluid migration can cause SP anomalies in highly permeable formations, as shown in figure Invasion is very shallow near the lower boundary of each permeable interval and deeper near the upper boundary.

The SP curve is rounded at the upper boundary because of the deep invasion, and it may have a sawtooth profile at thin, impervious shale streaks in which the SP deflection exceeds the SSP above the shale streak. A reading greater than the SSP is caused by the accumulation of filtrate below the shale streak. Encircling the hole is a horizontal disk of shale sandwiched between salt water and fresher mud filtrate that acts like a battery cell. The EMF of this cell is superimposed on the normal SSP, producing the sawtooth profile.

Noisy SP Logs

SP measuring circuits are sensitive and therefore prone to recording spurious electrical noise superimposed on the SP curve. Occasionally, the source of noise cannot be eliminated during logging, and a noisy log is recorded. However, this does not always render the log unusable.

A regular sine-wave signal may be superimposed on the SP curve when some part of the logging winch is magnetized. An intermittent contact between the casing and cable armor may also cause spurious spikes on the SP curve. In these situations, the SP curve can usually be read so that the sine-wave amplitude or noise spikes are not added to or subtracted from the authentic SP deflection.

Direct currents flowing through formations near the SP electrode can cause erroneous SP readings, particularly where formation resistivities are high. These currents may be caused by "bimetallism," when the logging tool has exposed metal housings. The currents are small and have a significant effect on the SP only in highly resistive formations. If an SP curve looks questionable in highly resistive formations, it should be relied on only in lower-resistivity intervals.

The offshore logging environment is notorious for its ample supply of sources of electrical noise, such as:

- Wave motion,

- Cathodic protection systems,

- Rig welding,

- Onboard generators,

- Leaky power sources.

On land, proximity to power lines and pumping wells may have a similar effect on the SP curve, but the effects can usually be minimized by carefully choosing the ground-electrode location.

Formation Evaluation Neutron Porosity

In the field of formation evaluation, porosity is one of the key measurements to quantify oil and gas reserves. Neutron porosity measurement employs a neutron source to measure the hydrogen index in a reservoir, which is directly related to porosity. The Hydrogen Index (HI) of a material is defined as the ratio of the concentration of hydrogen atoms per cm^3 in the material, to that of pure water at 75 °F. As hydrogen atoms are present in both water and oil filled reservoirs, measurement of the amount allows estimation of the amount of liquid-filled porosity.

Physics

Neutrons are typically emitted by a radioactive source such as Americium Beryllium (Am-Be) or Plutonium Beryllium (Pu-Be), or generated by electronic neutron generators such as minitron. Fast neutrons are emitted by these sources with energy ranges from 4 MeV to 14 MeV, and inelastically interact with matter. Once slowed down to 2 MeV, they start to scatter elastically and slow down further until the neutrons reach a thermal energy level of about 0.025 eV. When thermal neutrons are then absorbed, gamma rays are emitted. A suitable detector, positioned at a certain distance from the source, can measure either epithermal neutron population, thermal neutron population, or the gamma rays emitted after the absorption.

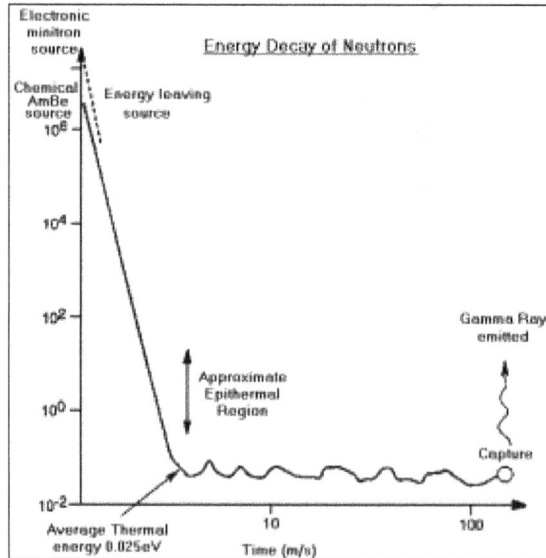

Neutron Energy Decay

Mechanics of elastic collisions predict that the maximum energy transfer occurs during collisions of two particles of equal mass. Therefore, a hydrogen atom (H) will cause a neutron to slow down the most, as they are of roughly equal mass. As hydrogen is fundamentally associated to the amount of water and oil present in the pore space, measurement of neutron population within the investigated volume is directly linked to porosity.

Correction

Determination of porosity is one of the most important uses of neutron porosity log. Correction parameters for lithology, borehole parameters, and others are necessary for accurate porosity determination as follow:

1.   Borehole size.

2. Borehole salinity.

3. Borehole temperature and pressure.

4. Mud cake.

5. Mud weight.

6. Formation salinity.

7. Tool standoff from borehole wall.

Interpretation

Subject to various assumptions and corrections, values of apparent porosity can be derived from any neutron log. One can not underestimate the slow down of neutrons by other elements even if they are less effective. Certain effects, such as lithology, clay content, and amount and type of hydrocarbons, can be recognized and corrected for only if additional porosity information is available, for example from sonic and density log. Any interpretation of a neutron log alone should be undertaken with a realization of the uncertainties involved.

Effect of Light Hydrocarbon and Gas

The quantitative response of neutron tool to gas or light hydrocarbon depends primarily on hydrogen index and "excavation effect". The hydrogen index can be estimated from the composition and density of the hydrocarbons.

Given a fixed volume, gas has considerably lower hydrogen concentration. When pore spaces in the rock are excavated and replaced with gas, the formation has smaller neutron-slowing characteristic, hence the terms "Excavation Effect". If this effect is ignored, a neutron log will show a low porosity value. This characteristic allows a neutron porosity log to be used with other porosity logs (such as a density log) to detect gas zones and identify gas-liquid contacts.

Measurement Technique

Neutron tools are based on the measurement of a neutron cloud of different energy levels within the investigated volume. Epithermal-neutron tools measure epithermal neutron density with energy levels between 100eV and 0.1eV in the formation. Thermal-neutron tools only measure the population of neutrons with a thermal energy level, and Neutron-gamma tools measure the intensity of gamma flux generated by thermal neutron capture. The tools usually have two detectors (or more) with different spacings from the source to produce ratio of count rates, which theoretically reduce borehole effects.

A Helium-3 (He-3) filled proportional counter is the most common epithermal and thermal neutron detector. Helium has a high neutron capture cross section and produces the following reaction when interacting with a neutron.

$$^3He + {}^1n \rightarrow {}^1H + {}^3H + 764keV \text{ energy}$$

To boost the charge produced by the interaction between Helium and a Neutron, a high voltage is applied to the anode of the counter. A high operating voltage is chosen to give enough gain for counting purposes. Most Helium-3 counters use a quench gas to stabilize high voltage performance and prevent run-away.

Well Logging

Well logging, also known as borehole logging is the practice of making a detailed record (a well log) of the geologic formations penetrated by a borehole. The log may be based either on visual inspection of samples brought to the surface (geological logs) or on physical measurements made by instruments lowered into the hole (geophysical logs). Some types of geophysical well logs can be done during any phase of a well's history: drilling, completing, producing, or abandoning. Well logging is performed in boreholes drilled for the oil and gas, groundwater, mineral and geothermal exploration, as well as part of environmental and geotechnical studies.

Wireline Logging

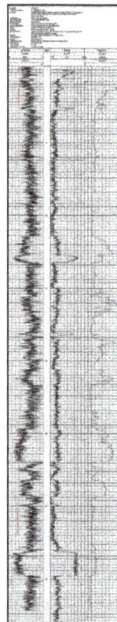

Wireline log consisting of caliper, density and resistivity logs.

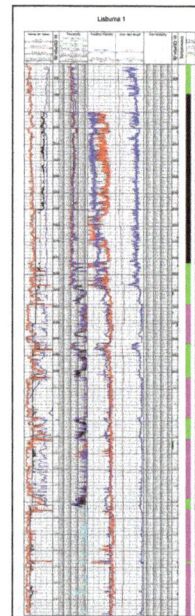

Wireline log consisting of a complete set of logs.

The oil and gas industry uses wireline logging to obtain a continuous record of a formation's rock properties. Wireline logging can be defined as being "The acquisition and analysis of geophysical data performed as a function of well bore depth, together with the provision of related services." Note that "wireline logging" and "mud logging" are not the same, yet are closely linked through the integration of the data sets. The measurements are made referenced to "TAH" - True Along Hole depth: these and the associated analysis can then be used to infer further properties, such as hydrocarbon saturation and formation pressure, and to make further drilling and production decisions.

Wireline logging is performed by lowering a 'logging tool' - or a string of one or more instruments - on the end of a wireline into an oil well (or borehole) and recording petrophysical properties using a variety of sensors. Logging tools developed over the years measure the natural gamma ray, electrical, acoustic, stimulated radioactive responses, electromagnetic, nuclear magnetic resonance, pressure and other properties of the rocks and their contained fluids.

The data itself is recorded either at surface (real-time mode), or in the hole (memory mode) to an electronic data format and then either a printed record or electronic presentation called a "well log" is provided to the client, along with an electronic copy of the raw data. Well logging operations can either be performed during the drilling process, to provide real-time information about the formations being penetrated by the borehole, or once the well has reached Total Depth and the whole depth of the borehole can be logged.

Real-time data is recorded directly against measured cable depth. Memory data is recorded against time, and then depth data is simultaneously measured against time. The two data sets are then merged using the common time base to create an instrument response versus depth log. Memory recorded depth can also be corrected in exactly the same way as real-time corrections are made, so there should be no difference in the attainable TAH accuracy.

The measured cable depth can be derived from a number of different measurements, but is usually either recorded based on a calibrated wheel counter, or (more accurately) using magnetic marks which provide calibrated increments of cable length. The measurements made must then be corrected for elastic stretch and temperature.

There are many types of wireline logs and they can be categorized either by their function or by the technology that they use. "Open hole logs" are run before the oil or gas well is lined with pipe or cased. "Cased hole logs" are run after the well is lined with casing or production pipe.

Wireline logs can be divided into broad categories based on the physical properties measured.

Conrad and Marcel Schlumberger, who founded Schlumberger Limited in 1926, are considered the inventors of electric well logging. Conrad developed the Schlumberger array, which was a technique for prospecting for metal ore deposits, and the brothers adapted that surface technique to subsurface applications. On September 5, 1927, a crew working for Schlumberger lowered an electric sonde or tool down a well in Pechelbronn, Alsace, France creating the first well log. In modern terms, the first log was a resistivity log that could be described as 3.5-meter upside-down lateral log.

In 1931, Henri George Doll and G. Dechatre, working for Schlumberger, discovered that the galvanometer wiggled even when no current was being passed through the logging cables down in the well. This led to the discovery of the spontaneous potential (SP) which was as important as the ability to measure resistivity. The SP effect was produced naturally by the borehole mud at the boundaries of permeable beds. By simultaneously recording SP and resistivity, loggers could distinguish between permeable oil-bearing beds and impermeable nonproducing beds.

In 1940, Schlumberger invented the spontaneous potential dipmeter; this instrument allowed the calculation of the dip and direction of the dip of a layer. The basic dipmeter was later enhanced by the resistivity dipmeter and the continuous resistivity dipmeter.

Oil-based mud (OBM) was first used in Rangely Field, Colorado in 1948. Normal electric logs require a conductive or water-based mud, but OBMs are nonconductive. The solution to this problem was the induction log, developed in the late 1940s.

The introduction of the transistor and integrated circuits in the 1960s made electric logs vastly more reliable. Computerization allowed much faster log processing, and dramatically expanded log data-gathering capacity. The 1970s brought more logs and computers. These included combo type logs where resistivity logs and porosity logs were recorded in one pass in the borehole.

The two types of porosity logs (acoustic logs and nuclear logs) date originally from the 1940s. Sonic logs grew out of technology developed during World War II. Nuclear logging has supplemented acoustic logging, but acoustic or sonic logs are still run on some combination logging tools.

Nuclear logging was initially developed to measure the natural gamma radiation emitted by underground formations. However, the industry quickly moved to logs that actively bombard rocks with nuclear particles. The gamma ray log, measuring the natural radioactivity, was introduced by Well Surveys Inc. in 1939, and the WSI neutron log came in 1941. The gamma ray log is particularly useful as shale beds which often provide a relatively low permeability cap over hydrocarbon reservoirs usually display a higher level of gamma radiation. These logs were important because they can be used in cased wells (wells with production casing). WSI quickly became part of Lane-Wells. During World War II, the US Government gave a near wartime monopoly on open-hole logging to Schlumberger, and a monopoly on cased-hole logging to Lane-Wells. Nuclear logs continued to evolve after the war.

After the discovery of nuclear magnetic resonance by Bloch and Purcell in 1946, the nuclear magnetic resonance log using the Earth's field was developed in the early 1950s by Chevron and Schlumberger. The NMR log was a scientific success but an engineering failure. More recent engineering developments by NUMAR (a subsidiary of Halliburton) in the 1990s has resulted in continuous NMR logging technology which is now applied in the oil and gas, water and metal exploration industry.

Many modern oil and gas wells are drilled directionally. At first, loggers had to run their tools somehow attached to the drill pipe if the well was not vertical. Modern techniques now permit continuous information at the surface. This is known as logging while drilling (LWD) or measurement-while-drilling (MWD). MWD logs use mud pulse technology to transmit data from the tools on the bottom of the drillstring to the processors at the surface.

Electrical Logs

Resistivity Log

Resistivity logging measures the subsurface electrical resistivity, which is the ability to impede the flow of electric current. This helps to differentiate between formations filled with salty waters (good conductors of electricity) and those filled with hydrocarbons (poor conductors of electricity). Resistivity and porosity measurements are used to calculate water saturation. Resistivity is expressed in ohms or ohms/meter, and is frequently charted on a logarithm scale versus depth because of the large range of resistivity. The distance from the borehole penetrated by the current varies with the tool, from a few centimeters to one meter.

Borehole Imaging

The term "borehole imaging" refers to those logging and data-processing methods that are used to produce centimeter-scale images of the borehole wall and the rocks that make it up. The context is, therefore, that of open hole, but some of the tools are closely related to their cased-hole equivalents. Borehole imaging has been one of the most rapidly advancing technologies in wireline well logging. The applications range from detailed reservoir description through reservoir performance to enhanced hydrocarbon recovery. Specific applications are fracture identification, analysis of small-scale sedimentological features, evaluation of net pay in thinly bedded formations, and the identification of breakouts (irregularities in the borehole wall that are aligned with the minimum horizontal stress and appear where stresses around the wellbore exceed the compressive strength of the rock). The subject area can be classified into four parts:

1. Optical imaging.

2. Acoustic imaging.

3. Electrical imaging.

4. Methods that draw on both acoustic and electrical imaging techniques using the same logging tool.

Porosity Logs

Porosity logs measure the fraction or percentage of pore volume in a volume of rock. Most porosity logs use either acoustic or nuclear technology. Acoustic logs measure characteristics of sound waves propagated through the well-bore environment. Nuclear logs utilize nuclear reactions that take place in the downhole logging instrument or in the formation. Nuclear logs include density logs and neutron logs, as well as gamma ray logs which are used for correlation. The basic principle behind the use of nuclear technology is that a neutron source placed near the formation whose porosity is being measured will result in neutrons being scattered by the hydrogen atoms, largely those present in the formation fluid. Since there is little difference in the neutrons scattered by hydrocarbons or water, the porosity measured gives a figure close to the true physical porosity whereas the figure obtained from electrical resistivity measurements is that due to the conductive formation fluid. The difference between neutron porosity and electrical porosity measurements therefore indicates the presence of hydrocarbons in the formation fluid.

Density

The density log measures the bulk density of a formation by bombarding it with a radioactive source and measuring the resulting gamma ray count after the effects of Compton Scattering and Photoelectric absorption. This bulk density can then be used to determine porosity.

Neutron Porosity

The neutron porosity log works by bombarding a formation with high energy epithermal neutrons that lose energy through elastic scattering to near thermal levels before being absorbed by the nuclei of the formation atoms. Depending on the particular type of neutron logging tool, either the gamma ray of capture, scattered thermal neutrons or scattered, higher energy epithermal neutrons are detected. The neutron porosity log is predominantly sensitive to the quantity of hydrogen atoms in a particular formation, which generally corresponds to rock porosity.

Boron is known to cause anomalously low neutron tool count rates due to it having a high capture cross section for thermal neutron absorption. An increase in hydrogen concentration in clay minerals has a similar effect on the count rate.

Sonic

A sonic log provides a formation interval transit time, which typically a function of lithology and rock texture but particularly porosity. The logging tool consists of a piezoelectric transmitter and receiver and the time taken to for the sound wave to travel the fixed distance between the two is recorded as an *interval transit time*.

Lithology Logs

Gamma Ray

A log of the natural radioactivity of the formation along the borehole, measured in API units, particularly useful for distinguishing between sands and shales in a siliclastic environment. This is because sandstones are usually nonradioactive quartz, whereas shales are naturally radioactive due to potassium isotopes in clays, and adsorbed uranium and thorium.

In some rocks, and in particular in carbonate rocks, the contribution from uranium can be large and erratic, and can cause the carbonate to be mistaken for a shale. In this case, the carbonate gamma ray is a better indicator of shaliness. The carbonate gamma ray log is a gamma ray log from which the uranium contribution has been subtracted.

Self/Spontaneous Potential

The Spontaneous Potential (SP) log measures the natural or spontaneous potential difference between the borehole and the surface, without any applied current. It was one of the first wireline logs to be developed, found when a single potential electrode was lowered into a well and a potential was measured relative to a fixed reference electrode at the surface.

The most useful component of this potential difference is the electrochemical potential because it can cause a significant deflection in the SP response opposite permeable beds. The magnitude of this deflection depends mainly on the salinity contrast between the drilling mud and the formation water, and the clay content of the permeable bed. Therefore, the SP log is commonly used to detect permeable beds and to estimate clay content and formation water salinity. The SP log can be used to distinguish between impermeable shale and permeable shale and porous sands.

Miscellaneous

Caliper

A tool that measures the diameter of the borehole, using either 2 or 4 arms. It can be used to detect regions where the borehole walls are compromised and the well logs may be less reliable.

Nuclear Magnetic Resonance

Nuclear magnetic resonance (NMR) logging uses the NMR response of a formation to directly determine its porosity and permeability, providing a continuous record along the length of the borehole. The chief application of the NMR tool is to determine moveable fluid volume (BVM) of a rock. This is the pore space excluding clay bound water (CBW) and irreducible water (BVI). Neither of these are moveable in the NMR sense, so these volumes are not easily observed on older logs. On modern tools, both CBW and BVI can often be seen in the signal response after transforming the relaxation curve to the porosity domain. Note that some of the moveable fluids (BVM) in the NMR sense are not actually moveable in the oilfield sense of the word. Residual oil and gas, heavy oil, and bitumen may appear moveable to the NMR precession measurement, but these will not necessarily flow into a well bore.

Spectral Noise Logging

Spectral noise logging (SNL) is an acoustic noise measuring technique used in oil and gas wells for well integrity analysis, identification of production and injection intervals and hydrodynamic characterisation of the reservoir. SNL records acoustic noise generated by fluid or gas flow through the reservoir or leaks in downhole well components.

Noise logging tools have been used in the petroleum industry for several decades. As far back as 1955, an acoustic detector was proposed for use in well integrity analysis to identify casing holes. Over many years, downhole noise logging tools proved effective in inflow and injectivity profiling of operating wells, leak detection, location of cross-flows behind casing, and even in determining reservoir fluid compositions. Robinson described how noise logging can be used to determine effective reservoir thickness.

Logging while Drilling

In the 1970s, a new approach to wireline logging was introduced in the form of logging while drilling (LWD). This technique provides similar well information to conventional wireline logging but instead of sensors being lowered into the well at the end of wireline cable, the sensors are integrated into the drill string and the measurements are made in real-time, whilst the well is being drilled. This allows drilling engineers and geologists to quickly obtain information such as porosity, resistivity, hole direction and weight-on-bit and they can use this information to make immediate decisions about the future of the well and the direction of drilling.

In LWD, measured data is transmitted to the surface in real time via pressure pulses in the well's mud fluid column. This mud telemetry method provides a bandwidth of less than 10 bits per second, although, as drilling through rock is a fairly slow process, data compression techniques mean that this is an ample bandwidth for real-time delivery of information. A higher sample rate of data is recorded into memory and retrieved when the drillstring is withdrawn at bit changes. High-definition downhole and subsurface information is available through networked or wired drillpipe that deliver memory quality data in real time.

Corrosion Well Logging

Throughout the life of the wells, integrity controles of the steel and cemented column (casing and tubing) are performed using calipers and thickness gauges. These advanced technical methods use non destructive technologies as ultrasonic, electromagnetic and magnetic transducers.

Memory Log

This method of data acquisition involves recording the sensor data into a down hole memory, rather than transmitting "Real Time" to surface. There are some advantages and disadvantages to this memory option.

- The tools can be conveyed into wells where the trajectory is deviated or extended beyond the reach of conventional Electric Wireline cables. This can involve a combination of weight to strength ratio of the electric cable over this extended reach. In such cases the memory tools can be conveyed on Pipe or Coil Tubing.

- The type of sensors are limited in comparison to those used on Electric Line, and tend to be focussed on the cased hole,production stage of the well. Although there are now developed some memory "Open Hole" compact formation evaluation tool combinations. These tools can be deployed and carried downhole concealed internally in drill pipe to protect them from damage while running in the hole, and then "Pumped" out the end at depth to initiate logging. Other basic open hole formation evaluation memory tools are available for use in "Commodity" markets on slickline to reduce costs and operating time.

- In cased hole operation there is normally a "Slick Line" intervention unit. This uses a solid mechanical wire (0.072 - 0.125 inches in OD), to manipulate or otherwise carry out operations in the well bore completion system. Memory operations are often carried out on this Slickline conveyance in preference to mobilizing a full service Electric Wireline unit.

- Since the results are not known until returned to surface, any realtime well dynamic changes cannot be monitored real time. This limits the ability to modify or change the well down hole production conditions accurately during the memory logging by changing the surface production rates. Something that is often done in Electric Line operations.

- Failure during recording is not known until the memory tools are retrieved. This loss of data can be a major issue on large offshore (expensive) locations. On land locations (e.g. South Texas, US) where there is what is called a "Commodity" Oil service sector, where logging often is without the rig infrastructure. this is less problematic, and logs are often run again without issue.

Coring

An example of a granite core.

Coring is the process of obtaining an actual sample of a rock formation from the borehole. There are two main types of coring: 'full coring', in which a sample of rock is obtained using a specialised drill-bit as the borehole is first penetrating the formation and 'sidewall coring', in which multiple samples are obtained from the side of the borehole after it has penetrated through a formation. The main advantage of sidewall coring over full coring are that it is cheaper (drilling doesn't have to be stopped) and multiple samples can be easily acquired, with the main disadvantages being that there can be uncertainty in the depth at which the sample was acquired and the tool can fail to acquire the sample.

Mud Logging

Mud logs are well logs prepared by describing rock or soil cuttings brought to the surface by mud circulating in the borehole. In the oil industry they are usually prepared by a mud logging company contracted by the operating company. One parameter a typical mud log displays is the formation gas (gas units or ppm). "The gas recorder usually is scaled in terms of arbitrary gas units, which are defined differently by the various gas-detector manufactures. In practice, significance is placed only on relative changes in the gas concentrations detected". The current oil industry standard mud log normally includes real-time drilling parameters such as rate of penetration (ROP), lithology, gas hydrocarbons, flow line temperature (temperature of the drilling fluid) and chlorides but may also include mud weight, estimated pore pressure and corrected d-exponent (corrected drilling exponent) for a pressure pack log. Other information that is normally notated on a mud log include directional data (deviation surveys), weight on bit, rotary speed, pump pressure, pump rate, viscosity, drill bit info, casing shoe depths, formation tops, mud pump info, to name just a few.

Information use

In the oil industry, the well and mud logs are usually transferred in 'real time' to the operating company, which uses these logs to make operational decisions about the well, to correlate formation depths with surrounding wells, and to make interpretations about the quantity and quality of hydrocarbons present. Specialists involved in well log interpretation are called log analysts.

Fractional Distillation

Fractional distillation is the process by which oil refineries separate crude oil into different, more useful hydrocarbon products based on their relative molecular weights in a distillation tower. This is the first step in the processing of crude oil, and it is considered to be the main separation process as it performs the initial rough separation of the different fuels. The different components that are separated out during this process are known as fractions. Fractions that are separated out include gasoline, diesel, kerosene, and bitumen.

Process

The process of fractional distillation is fairly simple, but is powerful in the way that it separates all the different, complex components of crude oil. First, the crude oil is heated to vapourize it and is fed into the bottom of a distillation tower. The resulting vapour then rises through the vertical column. As the gases rise through the tower, the temperature decreases. As the temperature decreases, certain hydrocarbons begin to condense and run off at different levels. Each fraction that condenses off at a certain level contains hydrocarbon molecules with a similar number of carbon atoms. These boiling point 'cuts' allow several hydrocarbons to be separated out in a single process. It is this cooling with the height of the tower that allows for the separation.

After this rough refinement, individual fuels may undergo more refinement to remove any contaminants or undesirable substances, or to improve the quality of the fuel through cracking.

Fractions

There are several ways of classifying the useful fractions that are distilled from crude oil. One general way is by dividing into three categories: light, middle, and heavy fractions. Heavier components condense at higher temperatures and are removed at the bottom of the column. The lighter fractions are able to rise higher in the column before they are cooled to their condensing temperature, allowing them to be removed at slightly higher levels. In addition to this, the fractions have the following properties:

- Light distillate is one of the more important fractions, and its products have boiling points around 70-200 °C. Useful hydrocarbons in this range include gasoline, naphta (a chemical feedstock), kerosene, jet fuel, and paraffin. These products are highly volatile, have small molecules, have low boiling points, flow easily, and ignite easily.

- Medium distillate are products that have boiling points of 200-350 °C. Products in this range include diesel fueland gas oil - used in the manufacturing of town gas and for commercial heating.

- Heavy distillate are the products with the lowest volatility and have boiling points above 350 °C. These fractions can be solid or semi-solid and may need to be heated in order to flow. Fuel oil is produced in this fraction. These products have large molecules, a low volatility, flow poorly, and do not ignite easily.

However, there are two major components that are not accounted for in these three categories. At the very top of the tower are the gases that are too volatile to condense,such as propane and butane. At the bottom are the "residuals" that contain heavy tars too dense to rise up the tower, including bitumen and other waxes. To further distill these they undergo steam or vacuum distillation as they are very useful.

Gamma Ray Logging

Gamma ray logging is a method of measuring naturally occurring gamma radiation to characterize the rock or sediment in a borehole or drill hole. It is a wireline logging method used in mining, mineral exploration, water-well drilling, for formation evaluation in oil and gas well drilling and for other related purposes. Different types of rock emit different amounts and different spectra of natural gamma radiation. In particular, shales usually emit more gamma rays than other sedimentary rocks, such as sandstone, gypsum, salt, coal, dolomite, or limestone because radioactive potassium is a common component in their clay content, and because the cation exchange capacity of clay causes them to adsorb uranium and thorium. This difference in radioactivity between shales and sandstones/carbonate rocks allows the gamma ray tool to distinguish between shales and non-shales. But it cannot distinguish between carbonates and sandstone as they both have similar deflections on the gamma ray log. Thus gamma ray logs cannot be said to make good lithological logs by themselves, but in practice, gamma ray logs are compared side-by-side with stratigraphic logs.

The gamma ray log, like other types of well logging, is done by lowering an instrument down the drill hole and recording gamma radiation variation with depth. In the United States, the device most commonly records measurements at 1/2-foot intervals. Gamma radiation is usually recorded in API units, a measurement originated by the petroleum industry. Gamma rays attenuate according to the diameter of the borehole mainly because of the properties of the fluid filling the borehole, but because gamma logs are generally used in a qualitative way, amplitude corrections are usually not necessary.

Three elements and their decay chains are responsible for the radiation emitted by rock: potassium, thorium and uranium. Shales often contain potassium as part of their clay content and tend to adsorb uranium and thorium as well. A common gamma-ray log records the total radiation and cannot distinguish between the radioactive elements, while a spectral gamma ray log can.

For standard gamma-ray logs, the measured value of gamma-ray radiation is calculated from concentration of uranium in ppm, thorium in ppm, and potassium in weight percent: e.g., GR API = 8 × uranium concentration in ppm + 4 × thorium concentration in ppm + 16 × potassium concentration in weight percent. Due to the weighted nature of uranium concentration in the GR API calculation, anomalous concentrations of uranium can cause clean sand reservoirs to appear shaley. For this reason, spectral gamma ray is used to provide an individual reading for each element so that anomalous concentrations can be found and properly interpreted.

An advantage of the gamma log over some other types of well logs is that it works through the steel and cement walls of cased boreholes. Although concrete and steel absorb some of the gamma radiation, enough travels through the steel and cement to allow for qualitative determinations.

In some places, non-shales exhibit elevated levels of gamma radiation. For instance, sandstones can contain uranium minerals, potassium feldspar, clay filling, or lithic fragments that cause the rock to have higher than usual gamma readings. Coal and dolomite may contain adsorbed uranium. Evaporite deposits may contain potassium minerals such as sylvite and carnallite. When this is the case, spectral gamma ray logging should be done to identify the source of these anomalies.

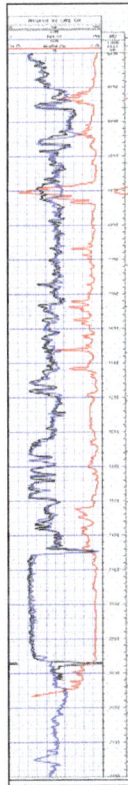

Example gamma ray log. Blue and black lines indicate the measured gamma rays. Sand section of interest is located at bottom of log where the log moves to the left.

Spectral Logging

Spectral logging is the technique of measuring the spectrum, or number and energy, of gamma rays emitted via natural radioactivity of the rock formation. There are three main sources of natural radioactivity on Earth: potassium (40K), thorium (principally 232Th and 230Th), and uranium (principally 238U and 235U). These radioactive isotopes each emit gamma rays that have a characteristic energy level measured in MeV. The quantity and energy of these gamma rays can be measured by a scintillometer. A log of the spectroscopic response to natural gamma ray radiation is usually presented as a total gamma ray log that plots the weight fraction of potassium (%), thorium (ppm) and uranium (ppm). The primary standards for the weight fractions are geological formations with known quantities of the three isotopes. Natural gamma ray spectroscopy logs became routinely used in the early 1970s, although they had been studied from the 1950s.

The characteristic gamma ray line that is associated with each radioactive component:

- Potassium: Gamma ray energy 1.46 MeV.

- Thorium series: Gamma ray energy 2.61 MeV.

- Uranium-Radium series: Gamma ray energy 1.76 MeV.

Another example of the use of spectral gamma ray logs is to identify specific clay types, like kaolinite or illite. This may be useful for interpreting the environment of deposition as kaolinite can form from feldspars in tropical soils by leaching of potassium; and low potassium readings may thus indicate the presence of one or more paleosols. The identification of specific clay minerals is also useful for calculating the effective porosity of reservoir rock.

Use in Mineral Exploration

Gamma ray logs are also used in mineral exploration, especially exploration for phosphates, uranium, and potassium salts.

Deepwater Drilling

Deepwater drilling, or Deep well drilling, is the process of creating holes by drilling rig for oil mining in deep sea. There are approximately 3400 deepwater wells in the Gulf of Mexico with depths greater than 150 meters.

It has not been technologically and economically feasible for many years, but with rising oil prices, more companies are investing in this area.

Recent industry analysis has estimated that the total expenditure in the global deepwater infrastructure market would reach $145bn in 2011.

"Not all oil is accessible on land or in shallow water. You can find some oil deposits buried deep under the ocean floor." Using sonic equipment, oil companies determine the drilling sites most likely to produce oil. Then they use a mobile offshore drilling unit (MODU) to dig the initial well. Some units are converted into production rigs, meaning they switch from drilling for oil to capturing oil once it's found. Most of the time, the oil company will replace the MODU with a more permanent oil production rig to capture oil. "The MODU's job is to drill down into the ocean's floor to find oil deposits. The part of the drill that extends below the deck and through the water is called the riser. The riser allows for drilling fluids to move between the floor and the rig. Engineers lower a drill string – a series of pipes designed to drill down to the oil deposit – through the riser."

"The expansion of deepwater drilling is happening despite accidents in offshore fields." In the Deepwater Horizon oil spill of 2010, a large explosion occurred killing workers and spilling oil into the Gulf of Mexico while a BP oil rig was drilling in deep waters.

"Despite the risks, the deepwater drilling trend is spreading in the Mediterranean and off the coast of East Africa after a string of huge discoveries of natural gas."

"The reason for the resumption of such drilling, analysts say, is continuing high demand for energy worldwide."

A Chinese ceramic model of a well with a water pulley system, excavated from a tomb of the Han Dynasty period.

Some of the earliest evidence of water wells are located in China. The Chinese discovered and made extensive use of deep drilled groundwater for drinking. The Chinese text *The Book of Changes*, originally a divination text of the Western Zhou dynasty, contains an entry describing how the ancient Chinese maintained their wells and protected their sources of water. Archaeological evidence and old Chinese documents reveal that the prehistoric and ancient Chinese had the aptitude and skills for digging deep water wells for drinking water as early as 6000 to 7000 years ago.

A well excavated at the Hemedu excavation site was believed to have been built during the Neolithic era. The well was cased by four rows of logs with a square frame attached to them at the top of the well. 60 additional tile wells southwest of Beijing are also believed to have been built around 600 BC for drinking and irrigation.

Types of Deepwater Drilling Facilities

Drilling in deep waters can be performed by two main types of mobile deepwater drilling rigs: semi-submersible drilling rigs and drillships. Drilling can also be performed from a fixed-position installation such as a fixed platform, or a floating platform, such as a spar platform, a tension-leg platform, or a semi-submersible production platform.

1. Fixed Platform - A Fixed Platform consists of a tall, (usually) steel structure that supports a deck. Because the Fixed Platform is anchored to the sea floor, it is very costly to build. This type of platform can be installed in water depth up to 500 meters (1,600 feet).

2. Jack-Up Rig - Jack-up rigs are mobile units with a floating hull that can be moved around; once arrived to the desired location, the legs are lowered to the sea floor and locked into place. Then the platform is raised up out of the water. That makes this type of rig safer to work on, because weather and waves are not an issue.

3. Compliant Tower Platform - A compliant tower is a particular type of fixed platform. Both are anchored to the sea floor and both work places are above the water surface. However, the compliant tower is taller and narrower, and can operate up to 1 kilometer (3,000 feet) water depth.

4. Semi-Submersible Production Platform - This platform is buoyant, meaning the bulk of it is floating above the surface. However, the well head is typically located on the sea floor, so extra precautions must be made to prevent a leak. A contributing cause to the oil spill disaster of 2010 was a failure of the leak-preventing system. These rigs can operate anywhere from 200 to 2,000 meters (660 to 6,560 feet) below the surface.

5. Tension-Leg Platform - The Tension-leg Platform consists of a floating structure, held in place by tendons that run down to the sea floor. These rigs drill smaller deposits in narrower areas, meaning this is a low-cost way to get a little oil, which attracts many companies. These rigs can drill anywhere from 200 to 1,200 meters (660 to 3,940 feet) below the surface.

6. Subsea System - Subsea Systems are actually wellheads, which sit on the sea floor and extract oil straight from the ground. They use pipes to force the oil back up to the surface, and can siphon oil to nearby platform rigs, a ship overhead, a local production hub, or even a faraway onshore site. This makes the Subsea system very versatile and a popular choice for companies.

7. Spar Platform - Spar Platforms use a large cylinder to support the floating deck from the sea floor. On average, about 90% of the Spar Platform's structure is underwater. Most Spar Platforms are used up to depths of 1 kilometer (3,000 feet), but new technology can extend them to function up to 3,500 meters (11,500 feet) below the surface. That makes it one of the deepest drilling rigs in use today.

Directional Drilling

Directional drilling (or slant drilling) is the practice of drilling non-vertical wells. It can be broken down into four main groups: oilfield directional drilling, utility installation directional drilling (horizontal directional drilling), directional boring, and surface in seam (SIS), which horizontally intersects a vertical well target to extract coal bed methane.

Many prerequisites enabled this suite of technologies to become productive. Probably, the first requirement was the realization that oil wells, or water wells, are not necessarily vertical. This realization was quite slow, and did not really grasp the attention of the oil industry until the late 1920s when there were several lawsuits alleging that wells drilled from a rig on one property had crossed the boundary and were penetrating a reservoir on an adjacent property. Initially, proxy evidence such as production changes in other wells was accepted, but such cases fueled the development of small diameter tools capable of surveying wells during drilling. Horizontal directional drill rigs are developing towards large-scale, micro-miniaturization, mechanical automation, hard stratum working, exceeding length and depth oriented monitored drilling.

A horizontal directional drill in operation.

Measuring the inclination of a wellbore (its deviation from the vertical) is comparatively simple, requiring only a pendulum. Measuring the azimuth (direction with respect to the geographic grid in which the wellbore was running from the vertical), however, was more difficult. In certain circumstances, magnetic fields could be used, but would be influenced by metalwork used inside wellbores, as well as the metalwork used in drilling equipment. The next advance was in the modification of small gyroscopic compasses by the Sperry Corporation, which was making similar compasses for aeronautical navigation.

Sperry did this under contract to Sun Oil (which was involved in a lawsuit as described above), and a spin-off company "Sperry Sun" was formed, which brand continues to this day, absorbed into Halliburton. Three components are measured at any given point in a wellbore in order to determine its position: the depth of the point along the course of the borehole (measured depth), the inclination at the point, and the magnetic azimuth at the point. These three components combined are referred to as a "survey". A series of consecutive surveys are needed to track the progress and location of a wellbore.

Prior experience with rotary drilling had established several principles for the configuration of drilling equipment down hole ("bottom hole assembly" or "BHA") that would be prone to "drilling crooked hole" (i.e., initial accidental deviations from the vertical would be increased). Counter-experience had also given early directional drillers ("DD's") principles of BHA design and drilling practice that would help bring a crooked hole nearer the vertical.

In 1934, H. John Eastman & Roman W. Hines of Long Beach, California, became pioneers in directional drilling when they and George Failing of Enid, Oklahoma, saved the Conroe, Texas, oil field. Failing had recently patented a portable drilling truck. He had started his company in 1931 when he mated a drilling rig to a truck and a power take-off assembly. The innovation allowed rapid drilling of a series of slanted wells. This capacity to quickly drill multiple relief wells and relieve the enormous gas pressure was critical to extinguishing the Conroe fire. In a May, 1934, *Popular Science Monthly* article, it was stated that "Only a handful of men in the world have the strange power to make a bit, rotating a mile below ground at the end of a steel drill pipe, snake its way in a curve or around a dog-leg angle, to reach a desired objective." Eastman Whipstock, Inc., would become the world's largest directional company in 1973.

Combined, these survey tools and BHA designs made directional drilling possible, but it was perceived as arcane. The next major advance was in the 1970s, when downhole drilling motors (aka mud motors, driven by the hydraulic power of drilling mud circulated down the drill string) became common. These allowed the drill bit to continue rotating at the cutting face at the bottom of the hole, while most of the drill pipe was held stationary. A piece of bent pipe (a "bent sub") between the stationary drill pipe and the top of the motor allowed the direction of the wellbore to be changed without needing to pull all the drill pipe out and place another whipstock. Coupled with the development of measurement while drilling tools (using mud pulse telemetry, networked or wired pipe or electromagnetism (EM) telemetry, which allows tools down hole to send directional data back to the surface without disturbing drilling operations), directional drilling became easier.

Certain profiles cannot be drilled while the drill pipe is rotating. Drilling directionally with a downhole motor requires occasionally stopping rotation of the drill pipe and "sliding" the pipe through the channel as the motor cuts a curved path. "Sliding" can be difficult in some formations, and it is almost always slower and therefore more expensive than drilling while the pipe is rotating, so the ability to steer the bit while the drill pipe is rotating is desirable. Several companies have developed tools which allow directional control while rotating. These tools are referred to as rotary steerable systems (RSS). RSS technology has made access and directional control possible in previously inaccessible or uncontrollable formations.

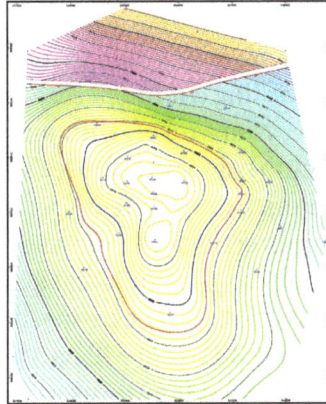

Structure map generated by contour map software for an 8,500-foot-deep (2,600 m) gas and oil reservoir in the Erath field, Vermilion Parish, Erath, Louisiana. The left-to-right gap, near the top of the contour map indicates a fault line. This fault line is between the blue/green contour lines and the purple/red/yellow contour lines. The thin red circular contour line in the middle of the map indicates the top of the oil reservoir. Because gas floats above oil, the thin red contour line marks the gas/oil contact zone. Directional drilling would be used to target the gas and oil reservoir.

Benefits

Wells are drilled directionally for several purposes:

- Increasing the exposed section length through the reservoir by drilling through the reservoir at an angle.

- Drilling into the reservoir where vertical access is difficult or not possible. For instance an oilfield under a town, under a lake, or underneath a difficult-to-drill formation.

- Allowing more wellheads to be grouped together on one surface location can allow fewer rig moves, less surface area disturbance, and make it easier and cheaper to complete and produce the wells. For instance, on an oil platform or jacket offshore, 40 or more wells can be grouped together. The wells will fan out from the platform into the reservoir(s) below. This concept is being applied to land wells, allowing multiple subsurface locations to be reached from one pad, reducing costs.

- Drilling along the underside of a reservoir-constraining fault allows multiple productive sands to be completed at the highest stratigraphic points.

- Drilling a "relief well" to relieve the pressure of a well producing without restraint (a "blowout"). In this scenario, another well could be drilled starting at a safe distance away from the blowout, but intersecting the troubled wellbore. Then, heavy fluid (kill fluid) is pumped into the relief wellbore to suppress the high pressure in the original wellbore causing the blowout.

Most directional drillers are given a blue well path to follow that is predetermined by engineers and geologists before the drilling commences. When the directional driller starts the drilling process, periodic surveys are taken with a downhole instrument to provide survey data (inclination and azimuth) of the well bore. These pictures are typically taken at intervals between 10–150 meters (30–500 feet), with 30 meters (90 feet) common during active changes of angle or direction, and distances of 60–100 meters (200–300 feet) being typical while "drilling ahead" (not making active changes to angle and direction). During critical angle and direction changes, especially while using a downhole motor, a measurement while drilling) (MWD) tool will be added to the drill string to provide continuously updated measurements that may be used for (near) real-time adjustments.

This data indicates if the well is following the planned path and whether the orientation of the drilling assembly is causing the well to deviate as planned. Corrections are regularly made by techniques as simple as adjusting rotation speed or the drill string weight (weight on bottom) and stiffness, as well as more complicated and time-consuming methods, such as introducing a downhole motor. Such pictures, or surveys, are plotted and maintained as an engineering and legal record describing the path of the well bore. The survey pictures taken while drilling are typically confirmed by a later survey in full of the borehole, typically using a "multi-shot camera" device.

The multi-shot camera advances the film at time intervals so that by dropping the camera instrument in a sealed tubular housing inside the drilling string (down to just above the drilling bit) and then withdrawing the drill string at time intervals, the well may be fully surveyed at regular depth intervals (approximately every 30 meters (90 feet) being common, the typical length of 2 or 3 joints of drill pipe, known as a stand, since most drilling rigs "stand back" the pipe withdrawn from the hole at such increments, known as "stands").

Drilling to targets far laterally from the surface location requires careful planning and design. The current record holders manage wells over 10 km (6.2 mi) away from the surface location at a true vertical depth (TVD) of only 1,600–2,600 m (5,200–8,500 ft).

This form of drilling can also reduce the environmental cost and scarring of the landscape. Previously, long lengths of landscape had to be removed from the surface. This is no longer required with directional drilling.

Disadvantages

Until the arrival of modern downhole motors and better tools to measure inclination and azimuth of the hole, directional drilling and horizontal drilling was much slower than vertical drilling due to the need to stop regularly and take time-consuming surveys, and due to slower progress in drilling itself (lower rate of penetration). These disadvantages have shrunk over time as downhole motors became more efficient and semi-continuous surveying became possible.

What remains is a difference in operating costs: for wells with an inclination of less than 40 degrees, tools to carry out adjustments or repair work can be lowered by gravity on cable into the hole. For higher inclinations, more expensive equipment has to be mobilized to push tools down the hole.

Another disadvantage of wells with a high inclination was that prevention of sand influx into the well was less reliable and needed higher effort. Again, this disadvantage has diminished such that, provided sand control is adequately planned, it is possible to carry it out reliably.

Stealing Oil

In 1990, Iraq accused Kuwait of stealing Iraq's oil through slant drilling. The United Nations redrew the border after the 1991 Gulf war, which ended the seven-month Iraqi occupation of Kuwait. As part of the reconstruction, 11 new oil wells were placed among the existing 600. Some farms and an old naval base that used to be in the Iraqi side became part of Kuwait.

In the mid-twentieth century, a slant-drilling scandal occurred in the huge East Texas Oil Field.

New Technologies

Between 1985 and 1993, the Naval Civil Engineering Laboratory (NCEL) (now the Naval Facilities Engineering Service Center (NFESC)) of Port Hueneme, California developed controllable horizontal drilling technologies. These technologies are capable of reaching 10,000–15,000 ft (3000–4500 m) and may reach 25,000 ft (7500 m) when used under favorable conditions.

Pore Pressure Gradient

Pore pressure gradient is a dimensional petrophysical term used by drilling engineers and mud engineers during the design of drilling programs for drilling (constructing) oil and gas wells into the earth. It is the pressure gradient inside the pore space of the rock column from the surface of the ground down to the total depth (TD), as compared to the pressure gradient of seawater in deep water.

Whereas in "pure math," the gradient of a scalar function expressed by the math notation grad(f) may not have physical units associated with it; in drilling engineering the pore pressure gradient is usually expressed in API-type International Association of Drilling Contractors (IADC) physical units of measurement, namely "psi per foot." In the well-known formula:

P = 0.052 * mud weight * true vertical depth

taught in almost all petroleum engineering courses worldwide, the mud weight (MW) is expressed in pounds per U.S. gallon, and the true vertical depth (TVD) is expressed in feet, and 0.052 is a commonly used conversion constant that can be derived by dimensional analysis:

$$\frac{1\,\text{psi}}{\text{ft}} \times \frac{1\,\text{ft}}{12\,\text{in}} \times \frac{1\,\text{lb/in}^2}{1\,\text{psi}} \times \frac{231\,\text{in}^3}{1\,\text{US Gal}} = 19.25000000\,\text{lb/gal}$$

It would be more accurate to divide a value in lb/gal by 19.25 than to multiply that value by 0.052. The magnitude of the error caused by multiplying by 0.052 is approximately 0.1%.

Example: For a column of fresh water of 8.33 pounds per gallon (lb/U.S. gal) standing still hydrostatically in a 21,000 feet vertical cased wellbore from top to bottom (vertical hole), the pressure gradient would be,

grad(P) = pressure gradient = 8.33 / 19.25 = 0.43273 psi/ft

and the hydrostatic bottom hole pressure (BHP) is then:

BHP = TVD * grad(P) = 21,000 * 0.43273 = 9,087 psi.

However, the formation fluid pressure (pore pressure) is usually much greater than a column of fresh water, and can be as much as 19 lb/U.S. gal (e.g., in Iran). For an onshore vertical wellbore with an exposed open hole interval at 21,000 feet with a pore pressure gradient of 19 lb/U.S. gal, the BHP would be:

BHP = pore pres grad * TVD = 21,000 * 19 / 19.25 = 20,727 psi

The calculation of a bottom hole pressure and the pressure induced by a static column of fluid are the most important and basic calculations in all well control courses taught worldwide for the prevention of oil and gas well blowouts.

Simple examples:

Using the figures above, we can calculate the maximum pressure at various depths in an offshore oil well.

Saltwater is 0.444 psi/ft (2.5% higher than fresh water but this not general and depends on salt concentration in water) Pore pressure in the rock could be as high as 1.0 psi/ft of depth (19.25 lb/gal).

A well with 5,000 feet of seawater and 15,000 feet of rock could have an overburden pressures at the bottom as high as 17,220 psi (5000 × 0.444 + 15000 × 1.0). That pressure is reduced at the surface by the weight of oil and gas the riser pipe, but this is only a small percentage of the total. It takes heavy mud (drilling fluid) inserted at the bottom to control the well when pressures are this high.

Pumpjack

A pumpjack is the overground drive for a reciprocating piston pump in an oil well.

It is used to mechanically lift liquid out of the well if not enough bottom hole pressure exists for the liquid to flow all the way to the surface. The arrangement is commonly used for onshore wells producing little oil. Pumpjacks are common in oil-rich areas.

Depending on the size of the pump, it generally produces 5 to 40 litres (1 to 9 imp gal; 1.5 to 10.5 US gal) of liquid at each stroke. Often this is an emulsion of crude oil and water. Pump size is also determined by the depth and weight of the oil to remove, with deeper extraction requiring more power to move the increased weight of the discharge column (discharge head).

A beam-type pumpjack converts the rotary motion of the motor to the vertical reciprocating motion necessary to drive the polished-rod and accompanying sucker rod and column (fluid) load. The engineering term for this type of mechanism is a walking beam. It was often employed in stationary and marine steam engine designs in the 18th and 19th centuries.

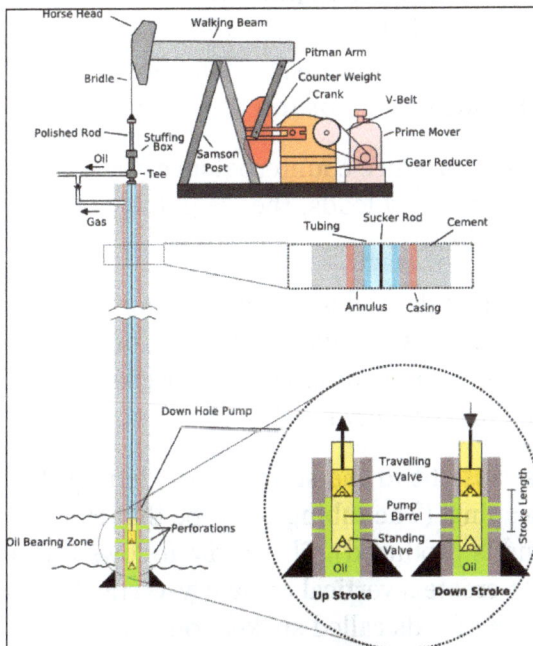

A diagram of a pumpjack.

Above Ground

In the early days, pumpjacks worked by rod lines running horizontally above the ground to a wheel on a rotating eccentric in a mechanism known as a central power. The central power, which might operate a dozen or more pumpjacks, would be powered by a steam or internal combustion engine or by an electric motor. Among the advantages of this scheme was only having one motor to power all the pumpjacks rather than individual motors for each. However, among the many difficulties was maintaining system balance as individual well loads changed.

Modern pumpjacks are powered by a prime mover. This is commonly an electric motor, but internal combustion engines are used in isolated locations without access to electricity, or, in the cases of water pumpjacks, where three-phase power is not available (while single phase motors exist at least up to 60 hp, providing power to single-phase motors above 10 horsepower can cause powerline problems, and many pumps require more than 10 horsepower). Common off-grid pumpjack engines run on natural gas, often casing gas produced from the well, but pumpjacks have been run on many types of fuel, such as propane and diesel fuel. In harsh climates, such motors and engines may be housed in a shack for protection from the elements. Engines that power water pumpjacks often receive natural gas from the nearest available gas grid.

The prime mover runs a set of pulleys to the transmission, often a double-reduction gearbox, which drives a pair of cranks, generally with counterweights installed on them to assist the motor in lifting the heavy rod assembly. The cranks raise and lower one end of an I-beam which is free to move on an A-frame. On the other end of the beam is a curved metal box called a horse head or donkey head, so named due to its appearance. A cable made of steel—occasionally, fibreglass—called a bridle, connects the horse head to the polished rod, a piston that passes through the stuffing box.

The cranks themselves also produce counterbalance due to their weight, so on pumpjacks that do not carry very heavy loads, the weight of the cranks themselves may be enough to balance the well load.

Sometimes, however, crank-balanced units can become prohibitively heavy due to the need for counterweights. Currently, Lufkin Industries offer "air-balanced" units, where counterbalance is provided by a pneumatic cylinder charged with air from a compressor, eliminating the need for counterweights.

The polished rod has a close fit to the stuffing box, letting it move in and out of the tubing without fluid escaping. (The tubing is a pipe that runs to the bottom of the well through which the liquid is produced). The bridle follows the curve of the horse head as it lowers and raises to create a vertical or nearly-vertical stroke. The polished rod is connected to a long string of rods called sucker rods, which run through the tubing to the down-hole pump, usually positioned near the bottom of the well.

The densely developed Kern River Oil Field, California: Hundreds of pumpjacks
are visible in the full-size view. This style of development was common in the
oil booms of the early 20th century.

Down-hole

At the bottom of the tubing is the down-hole pump. This pump has two ball check valves: a stationary valve at bottom called the standing valve, and a valve on the piston connected to the bottom of the sucker rods that travels up and down as the rods reciprocate, known as the traveling valve. Reservoir fluid enters from the formation into the bottom of the borehole through perforations that have been made through the casing and cement (the casing is a larger metal pipe that runs the length of the well, which has cement placed between it and the earth; the tubing, pump, and sucker rod are all inside the casing).

When the rods at the pump end are travelling up, the traveling valve is closed and the standing valve is open (due to the drop in pressure in the pump barrel). Consequently, the pump barrel fills with the fluid from the formation as the traveling piston lifts the previous contents of the barrel upwards. When the rods begin pushing down, the traveling valve opens and the standing valve closes (due to an increase in pressure in the pump barrel). The traveling valve drops through the fluid in the barrel (which had been sucked in during the upstroke). The piston then reaches the end of its stroke and begins its path upwards again, repeating the process.

Often, gas is produced through the same perforations as the oil. This can be problematic if gas enters the pump, because it can result in what is known as gas locking, where insufficient pressure builds up in the pump barrel to open the valves (due to compression of the gas) and little or nothing is pumped. To preclude this, the inlet for the pump can be placed below the perforations. As the gas-laden fluid enters the well bore through the perforations, the gas bubbles up the annulus (the space between the casing and the tubing) while the liquid moves down to the standing valve inlet. Once at the surface, the gas is collected through piping connected to the annulus.

Water Well Pumpjacks

Pumpjacks can also be used to drive what would now be considered old-fashioned hand-pumped water wells. The scale of the technology is frequently smaller than for an oil well, and can typically fit on top of an existing hand-pumped well head. The technology is simple, typically using a parallel-bar double-cam lift driven from a low-power electric motor, although the number of pumpjacks with stroke lengths 54 inches (137 cm) and longer being used as water pumps is increasing.

Although the flow rate for a water well pumpjack is lower than that from a jet pump and the lifted water is not pressurised, the beam pumping unit has the option of hand pumping in an emergency, by hand-rotating the pumpjack cam to its lowest position, and attaching a manual handle to the top of the wellhead rod. In larger pumpjacks powered by engines, the engine can run off fuel stored in a reservoir or from natural gas delivered from the nearest gas grid. In some cases, this type of pump consumes less power than a jet pump and is, therefore, cheaper to run.

Downhole Oil–Water Separation Technology

Downhole oil–water separation (DOWS) technology is an emerging technology that separates oil and gas from produced water at the bottom of the well, and re-injects most of the produced water into another formation which is usually deeper than the producing formation, while the oil and gas rich stream is pumped to the surface. DOWS effectively removes solids from the disposal fluid and thus avoids injectivity impairment caused by solids plugging. Simultaneous injection using DOWS minimizes the opportunity for the contamination of underground sources of drinking water (USDWs) through leaks in tubing and casing during the injection process.

Advantages

1. Lower cost of oil production: DOWS technology reduces more than 70% of the water that was supposed to be produced thereby reducing the total cost of lifting, treating, re-injecting and disposal of produced water, the cost of production of oil is greatly reduced.

2. Reduction of environmental impact of oil and gas operations: Standard injection well operations pose a significant but manageable risk of environmental pollution. The highest risks occur due to surface spills during re-injection of the separated brine. By reducing the amount of brine brought to the surface and re-injected the environmental risks associated with re-injection wells can be reduced.

3. Pollution of underground sources of drinking water: Produced water in most cases contains brines. With DOWS technology, dissolved impurities are not allowed to get to the surface hence they do not pass through intervals containing underground sources of drinking water (USDW). Also, since the water does not

require re-injection from the surface, it will not pass through intervals of USDWs then, either. Therefore, the risks posed by large volumes of these fluids passing USDWs both exiting and re-entering the well upon injection are minimized.

4. Because DOWS technology uses underground equipment, surface brine disposal operations, which may involve pumps, pipes, tank batteries, and other storage facilities may also be reduced in size and extent if not altogether eliminated.

5. The DOWS system is used for re-injection thereby beefing up the drawdown pressure which can result in increased production rate.

Drilling Mud

Drilling mud, also called drilling fluid, in petroleum engineering is a heavy, viscous fluid mixture that is used in oil and gas drilling operations to carry rock cuttings to the surface and also to lubricate and cool the drill bit. The drilling mud, by hydrostatic pressure, also helps prevent the collapse of unstable strata into the borehole and the intrusion of water from water-bearing strata that may be encountered.

Drilling muds are traditionally based on water, either fresh water, seawater, naturally occurring brines, or prepared brines. Many muds are oil-based, using direct products of petroleum refining such as diesel oil or mineral oil as the fluid matrix. In addition, various so-called synthetic-based muds are prepared using highly refined fluid compounds that are made to more-exacting property specifications than traditional petroleum-based oils. In general, water-based muds are satisfactory for the less-demanding drilling of conventional vertical wells at medium depths, whereas oil-based muds are better for greater depths or in directional or horizontal drilling, which place greater stress on the drilling apparatus. Synthetic-based muds were developed in response to environmental concerns over oil-based fluids, though all drilling muds are highly regulated in their composition, and in some cases specific combinations are banned from use in certain environments.

A typical water-based drilling mud contains a clay, usually bentonite, to give it enough viscosity to carry cutting chips to the surface, as well as a mineral such as barite (barium sulfate) to increase the weight of the column enough to stabilize the borehole. Smaller quantities of hundreds of other ingredients might be added, such as caustic soda (sodium hydroxide) to increase alkalinity and decrease corrosion, salts such as potassium chloride to reduce infiltration of water from the drilling fluid into the rock formation, and various petroleum-derived drilling lubricants. Oil- and synthetic-based muds contain water (usually a brine), bentonite and barite for viscosity and weight, and various emulsifiers and detergents for lubricity.

Drilling mud is pumped down the hollow drill pipe to the drill bit, where it exits the pipe and then is flushed back up the borehole to the surface. For economic and environmental reasons, oil- and synthetic-based muds are usually cleaned and recirculated (though some muds, particularly water-based muds, can be discharged into the surrounding environment in a regulated manner). Larger drill cuttings are removed by passing the returned mud through one or more vibrating screens, and sometimes fine cuttings are removed by passing the mud through centrifuges. Cleaned mud is blended with new mud for reuse down the borehole.

Drilling fluids are also employed in the drilling of water wells.

Mud Logging

Mud logging, in its conventional implementation, involves the rig-site monitoring and assessment of information that comes to the surface while drilling, with the exclusion of data from downhole sensors. The term mud logging is thought, by some, to be outdated and not sufficiently descriptive. Because of the relatively broad range of services performed by the geologists, engineers, and technicians traditionally called mud loggers, the term "surface logging" is sometimes used, and the personnel performing the services may be called surface-logging specialists. Additional specialist designations may include:

- Pore pressure engineer.

- Formation evaluation engineer.

- Logging geologist.

- Logging engineer.

For the sake of generality, the terms mud logging and mud logger are used here with the understanding that the hybrid discipline encompasses much more than monitoring the mud returns and that the trained specialists perform engineering and geological tasks that span several traditional disciplines.

Objectives of Mud Logging

There are several broad objectives targeted by mud logging: identify potentially productive hydrocarbon-bearing formations, identify marker or correlatable geological formations, and provide data to the driller that enables safe and economically optimized operations. The actions performed to accomplish these objectives include the following:

- Collecting drill cuttings.

- Describing the cuttings (type of minerals present).

- Interpreting the described cuttings (lithology).

- Estimating properties such as porosity and permeability of the drilled formation.

- Maintaining and monitoring drilling-related and safety-related sensing equipment.

- Estimating the pore pressure of the drilled formation.

- Collecting, monitoring, and evaluating hydrocarbons released from the drilled formations.

- Assessing the producibility of hydrocarbon-bearing formations.

- Maintaining a record of drilling parameters.

Mud logging service first focused on monitoring the drilling mud returns qualitatively for oil and gas content. This included watching the mud returns for oil sheen, monitoring the gas evolving from the mud as it depressured at the surface, and examining the drill cuttings to determine the rock type that had been drilled, as well as looking for indication of oil on the cuttings. Detection of the onset of abnormal formation pressures using drilling parameters was proposed with the introduction of the d exponent. Gas chromatography, which was developed early in the 20th century, saw its introduction in mud logging in the 1970s when electronics became sufficiently compact, rugged, and robust to be used at rig sites. The literature provides excellent reviews of the early history.

Computerized data acquisition and the ability to routinely transfer continuously acquired data to the office data center enabled the broader application of more sophisticated interpretive techniques and the integration of data from different sources into the geological and reservoir model, in near real time. This, coupled with the blossoming of measurement-while-drilling (MWD) and logging-while-drilling (LWD) tools, moved the mud-logging unit into a new role as a hub for rig-site data gathering and transmission. Starting in the 1980s, significant improvements to existing technologies, as well as major technical breakthroughs, have given the geologist and petroleum engineer a great number of powerful mud-logging tools to interpret and integrate geological, drilling, and geochemical data.

The traditional products delivered by a mud-logging vendor include:

- Geological evaluation.

- Petrophysical/reservoir formation evaluation.

- Drilling engineering support services.

In this overview, we consider that these products support three basic processes associated with drilling and evaluation of wells:

- Formation evaluation (building or refining the geological and reservoir models).

- Drilling engineering and operations (the planning and execution of the well construction process).

- Maintaining drilling and evaluation operations with appropriate health, safety, and environmental (HSE) consideration.

Mud Logging Data Acquisition

Figure schematically shows the components of a drilling operation that have a part in mud logging. The most critical component is the drilling fluid (drilling mud), which, in addition to its role in drilling mechanics, carries most of the information from the formation up to the surface where it is acquired, decoded, or extracted from the mud stream by various techniques. Drilling liberates gas and liquid formation fluids, and circulation of the drilling fluid carries these to the surface (except during riserless drilling in a deepwater offshore environment, in which the drilling returns circulate only up to the sea floor). Cuttings, pieces of formation rock, are also carried in the circulated drilling fluid. MWD and LWD data are frequently encoded as pressure pulses and transmitted to the surface. Mud temperature is not a direct indicator of subsurface formation temperature, but monitoring the trend is important to understanding gas extraction efficiency and recycling. In deepwater drilling environments, the mud can be cooled significantly on the trip from the sea floor to the surface.

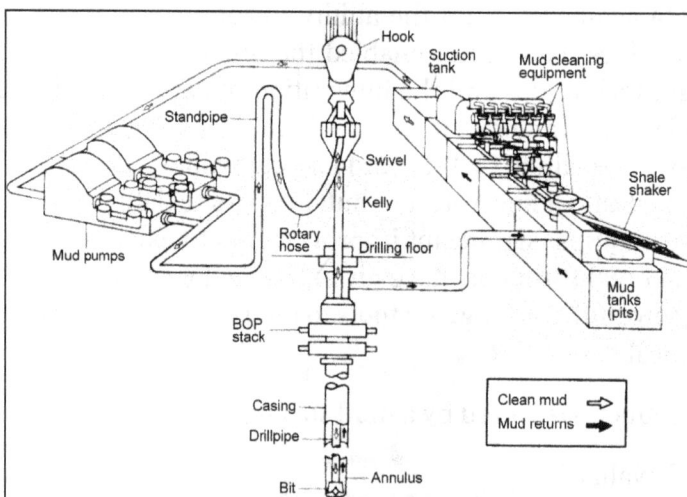

Drilling fluid flow path during drilling operations

Drilling fluid is stored in the mud pit, drawn into the mud pumps, and pumped into the drillpipe via the kelly. Mud travels down the drillpipe, through any MWD tools and drill motors, and through the bit nozzles where its discharge aids drilling mechanics. At this point, the drilling fluid carries away rock fragments from drilled formation, along with any liberated reservoir fluids (water, oil, or gas). Cuttings and reservoir fluids are transported to the surface. Any gaseous components are dissolved in the base fluid of the drilling mud under most overbalanced drilling conditions. Drilling mud continues its flow up the wellbore drillstring annulus, through the casing-drillpipe annulus and the blowout preventer (BOP) stack, and, in the case of an offshore well, up the riser. At the bell nipple, the returning drill fluid is exposed to atmospheric pressure and flows down the mud return line. If an underbalanced drilling operation is being used, there is a rotating seal around the drillpipe, and the pressurized drilling returns stream moves through the "blooey line" to a separator and flair.

The return mud stream continues down the return line to the shaker box or "possum belly." This is the standard location for the "gas trap" gas extractor. Mud pours over the shaker screens, with cuttings getting discharged off the top of the screen, while the drilling fluid that falls through the screens travels on through the degasser, desander, and desilter to the mud pits. The mud logger takes samples or acquires data at the following points in the process:

- Whole mud samples are taken at the mud suction pit and at the possum belly and are used to do whole-mud extraction using a steam still. They may be taken on an occasional basis during coring and wireline logging to assess the effects of mud filtrate and solids.

- Drill cuttings samples are taken off the shaker screen and off a "catch board" where cuttings fall from the screen to disposal. These are used for lithological and mineralogical description, paleo description, and sometimes "canned" for laboratory-based carbon-isotope analysis, detailed geological examinations such as thin-section preparation and analysis, chemostratigraphy, and source rock evaluation.

- Gas sampling is done through an extractor at the possum belly, in some cases at the bell nipple or off the mud return line to minimize losses to the atmosphere, and at the mud suction line or mud pit to monitor recycle gas content of the mud. After extraction, gas analysis may be performed at the sampling location, or, more routinely, the gas is continuously transferred via a vacuum line to the logging unit where it is passes through the manifold of analytical instruments (total HC, GC, MS, H_2S, etc) and may be captured for laboratory-based analysis (carbon isotope, molecular composition).

- Mud temperatures are monitored at the mud suction pit and mud return line.

The mud engineer collects mud samples for analyses that are used to determine any adjustments to mud properties needed for drilling.

Contamination is defined here as any material that does not come from the formation that has been drilled at the time that a specific volume element of mud exits the bit. Mud contamination has several potential sources:

- Air, which can enter the top of the drillpipe when the Kelly-drillpipe joint is broken during a connection.

- Pipe scale and pipe dope from inside the drillpipe (pipe dope fluoresces and may interfer with show identification or description).

- Rock sloughing or rubbing off formations further up hole.

- Cuttings that have bedded or built up because of improper hole cleaning dynamics that are mobilized by changes in mud viscosity, pumping rate, or drillpipe or collar rotation.

- Uphole fluids that flow or are swabbed into the annulus.

- Cuttings that have built up on the shaker screen or in the possum belly.

The logger should be watching for any change in cuttings or mud-conveyed hydrocarbon fluids that indicate contamination. Mud additives such as weighting agents and lost-circulation material are not considered contaminants, but must be monitored because some of these interfere with analytical observations and descriptions or give interfering instrument responses. Some base mud fluids, particularly some of the synthetic fluids, create challenges for the mud logger, as do some chemical additives (e.g., some sulfate or sulfonate wetting agents may give a false positive H_2S indication).

Samples of the drill cuttings are taken at the shale shaker. Wellsite geologists or engineers should specify the appropriate procedure for collecting samples, which may be done by the mud logger or the mud logger's sample catcher. Cuttings have a relatively short residence time on the shaker screen. Sampling protocol should include taking a composite sample with portions from different areas of the screen, combined with cuttings that have been retained on a "cuttings board." A cuttings board is a wooden board, steel angle iron, or other such device that is hung just below the base of the shaker screens to catch cuttings as they fall off the edge of the screen. Immediately after samples are collected, the screen and catch board should be washed down with clean drilling mud base fluid. The logger should mix this composite sample and take divided portions for cleaning, interpretation, and bagging. The wellsite planner should specify the sampling frequency (typically a composite over 10-, 30-, or 90-ft intervals or on a timed basis).

Gas sampling is traditionally done with a mechanical degasser, generically called a "gas trap." Figure shows an example. Typically placed in the shaker box, the trap pulls in drilling mud through the centrifugal action of the stirrer. The mechanical action of the stirrer, combined with a slight vacuum pulled in the trap head space, allows the gas to partition between the liquid and gas phase. The head-space gas is pulled by vacuum through tubing, into the logging unit, and on through the gas analysis manifold.

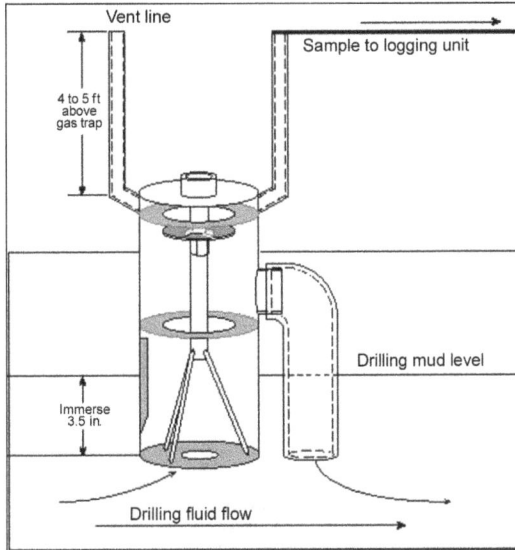

Schematic of a gas-trap type gas extractor.

Alternative methods for sampling gas may be accomplished by continuously operating controlled-volume mechanical or thermomechanical slip-stream gas extractors and membrane-type extractors. The mud logger may place the sampling point for these gas extraction devices at the bell nipple, the mud return line, or in the shaker box. Other methods require taking discrete samples, followed by thermal extraction techniques [such as the steam still, where samples of the whole mud are collected and portions heated in a steam-distillation apparatus] and microwave heating methods.

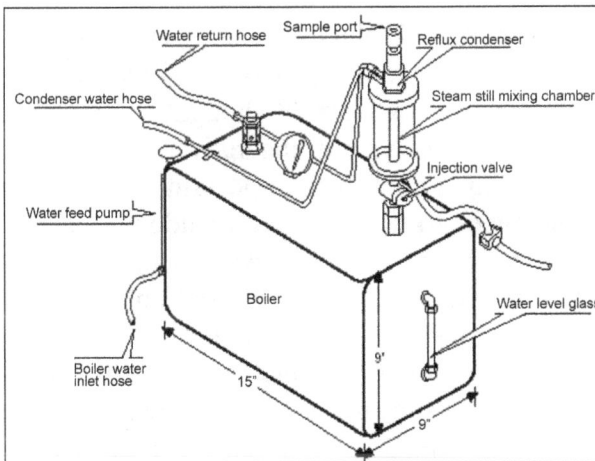

Schematic of a steam still-gas extractor.

The gas manifold may include provisions for a portion of the gas stream to be pumped into sample containers, either laminated gas bags or stainless steel tubes. These gas samples are then shipped from the rig for laboratory analyses. There are new mass-spectrometer-based techniques that may not require a bulk extraction of the gas from the mud for analysis.

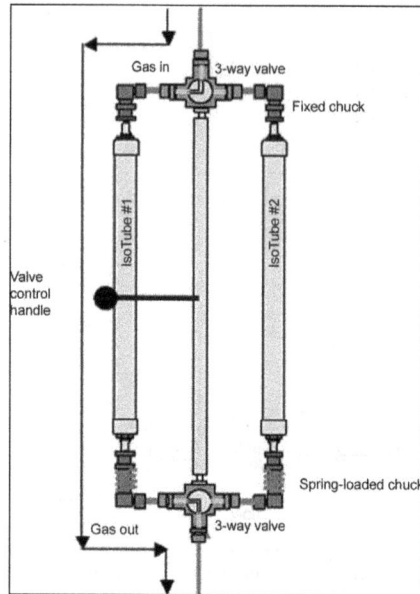

Low-pressure gas sampling tubes mounted two to a rack for continuous, sequential gas collection

Once gas is extracted from the drilling fluid, various analytical techniques determine properties of the gas at the rig site. The basic measurements include a determination of the "total" gas concentration and the composition and concentrations of the constituent components.

Maintaining Data Quality

Many different data are available, obtained through technologies that range from tried-and-true classical "wet chemistry" techniques through high-tech sensors that use procedures established after countless years of thoughtful research, development, and field testing. Even the best planned operations may, from time to time, provide data of poor quality or even totally miss data from important geologic intervals. Proper planning of surface data logging operations should include provisions for "whole-system" qualification before starting the operation, as well as a plan for the occasional quality audit. Details will vary widely depending on the location of the operations, availability of staff, project magnitude, and economics. An appropriate quality assurance program may be as simple as receiving a weekly e-mail or fax with GC calibration information or as intensive as scheduling rig-site audits, depending on the exact circumstances and well logging objectives.

Drilling Engineering and Operations

There is significant overlap between data gathered for geological, petrophysical, and reservoir engineering needs vs. data gathered for the driller. Information about pore pressure, formation gas, and rock type and strength are an integral part of well planning. Continuously tracking these parameters as a well is drilled and comparing the actual data with what was used in the well plan allows for quick response by the driller when trouble occurs. It also allows the driller to "fine tune" his operations to optimize drilling performance, which is measured by drill rate, trouble time and cost, and delivery of well specifications (e.g., in terms of being a producing asset, an exploration well, or an appraisal well). We will leave these items in the "evaluation" bin, with the acknowledgement that they could just as easily be lumped into the drilling engineering category.

Any measurable parameter that gives an indication of pore pressure provides the driller with an estimate of the degree of overbalance, which directly affects the rate of penetration (ROP). Mud weight will be adjusted to be within the desired window for a well's particular set of drilling dynamics and rock strengths on the basis of modeled drill rate.

Weight on Bit and Rate of Penetration

These data are collected to indicate drilling performance. The driller would like to know how to predict his drilling or penetration rate. Bourgoyne *et al.* describes several models that have been developed and used. Jorden proposed modifying the Bingham model and defined a normalized parameter called the drilling exponent (the *d*-exponent):

$$d_{exp} = \frac{\log\left(\dfrac{R}{60N}\right)}{\log\left(\dfrac{12W}{1,000d_b}\right)},$$

where R = the penetration rate in ft/hr, N = the rotary speed in rpm, W = the weight on bit in Mlbf, and d_b = the bit diameter in inches. The d-exponent is sometimes corrected for mud density changes by considering the effects of ρ_n, the mud density equivalent to a normal formation pore pressure, and ρ_e, the equivalent mud density at the bit while circulating:

$$d_c = d_{exp}\frac{\rho_n}{\rho_e}.$$

Mud Pit Level

Indicators specify changes in the volume of mud in the pit. The total volume of mud changes continuously with depth as the hole volume increases. Rapid increases in pit volume may mean an influx of reservoir fluids, and well control measures may need to be implemented. A rapid decrease in volume indicates a downhole loss of mud, and lost-circulation material will probably be added to the drilling fluid.

Mud Chloride Content

Mud chloride content is monitored in all systems, along with the water content in non-aqueous drilling fluid systems. Significant changes in content may indicate influx of formation water, which means that an underbalance condition may be close, and mud weight may need to be increased.

Lithology and Mineralogy

Lithology and mineralogy may change as a fault is approached. Warmer water, with higher concentrations of dissolved salts, can flow along faults during some phases of their development. As the water moves into cooler zones, salts will precipitate, plugging pores and showing up in the cuttings. Indication of an approaching fault may warn of a potential jump across the fault, which, in some areas, is accompanied by a significant change in pore pressure. Advance knowledge of this allows the driller to adjust the mud weight before he encounters problems.

Total Gas

Concentrations in the drilling fluid returns indicate the degree of overbalance or underbalance between the equivalent mud density and the formation pore pressure. The total gas concentration measured while drilling shales establishes a baseline or background level that is useful in tracking pore pressure, with the assumption that the shale pore fluids are in equilibrium with any neighboring permeable sands.

Figure indicates how trends of several parameters vary with pore pressure and depth. Monitoring and plotting these can give indications of the transition from normally pressured zones to geopressures. The ROP, drilling exponent, shale cutting density, and background total gas all follow a normal trend with depth. Attempts to calibrate these measurements directly to pore pressure have been somewhat successful and usually are on the basis of establishing the trend for normally pressured formations. When a deviation from the normal trend occurs, correlations specific to the basin or geographic region are used to estimate the formation pore pressure. Most logging companies offer pore pressure service, which requires experienced pore pressure engineers who, frequently through experience, add subjective input to the model as well as the objective parametric inputs.

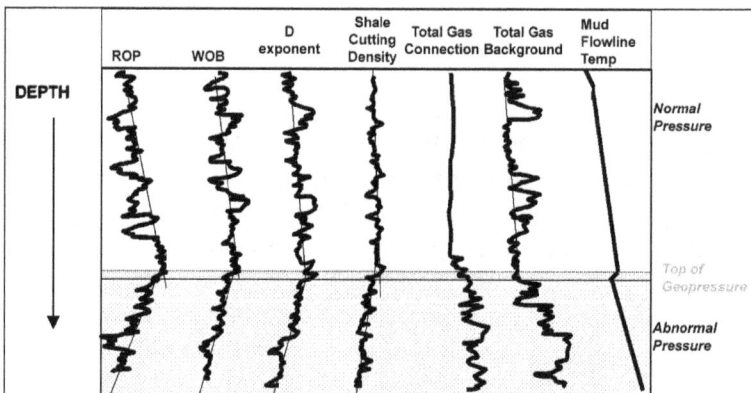

Schematic description of drilling and mud logging parameter
changes with depth, normal pressure trends, and geopressure trends.

While the accuracy of these particular methods will vary from site to site, such plots are extremely useful in identifying the transition into geopressures (i.e., when passing from normally pressured to abnormally pressured zones). At the transition to geopressures, the trend lines change slope. Because some of the changes may be subtle, looking at all the available data helps pinpoint the transition.

Connection Gas

Connection gas is a good indicator of swabbing the wellbore at the bit (i.e., reducing the mud pressure at the bottom of the hole to below the pore pressure). If the pore pressure is less than the swabbed bottomhole pressure, little or no connection gas is seen. Some knowledge of the dynamic rheology of the drilling fluid is needed to perform input into a "swab model."

Normal Geothermal Gradient

The normal geothermal gradient may shift on transition into geopressures. Other thermal nomalies, such as proximity to subsurface salt bodies, may interfer with this phenomenon. A more thorough discussion of these techniques and their application to detect overpressure may be found in several references.

Monitoring the Rate of Cuttings Return

As a formation is drilled, the cuttings should be circulated to the surface. Improper hole cleaning results in the downhole retention of cuttings, frequently as a cuttings bed on the low side of the hole in inclined wells. This causes an increased drag on the drillpipe and, if buildup is severe, may pack off the drillpipe, causing it to stick. Monitoring the rate of cuttings production from the drilling fluid returns indicates an approaching problem and warns the driller, which allows for remedial action before the pipe sticks. Watching for an increase in cuttings return rate can flag sloughing or extruding shale conditions, which call for an adjustment of mud density, as well as extreme washout conditions. Naegel et al. describe a device for continuously weighing the cuttings as they come off the shaker screens and comparing this with what would be expected for a given ROP and mud pump rate.

Health, Safety and Environmental Considerations

Various parameters measured for formation evaluation and to monitor drilling operations and equipment are also indicators of conditions that could pose health, safety, and environmental concerns. Pore pressure changes that result in loss of well control pose obvious safety concerns. Any loss of control that results in a hydrocarbon release also poses serious environmental issues. Ambient monitoring for natural gas is done for health and fire safety. Monitoring hydrogen sulfide (H_2S) is essential in areas in which the potential has been shown historically to exist, as well as in rank wildcat wells in which the characteristics of the geological basin are poorly known.

Hydrogen sulfide is detectable by GC but can not be measured with an FID. Thermal conductivity, MS, and solid state sensors detect H_2S. The Delphian Mud Duck, which uses an electrochemical sensor, monitors dissolved H_2S, HS-, and S^{2-} ionic concentrations to give total sulfide content of the drilling fluid. This tool continuously follows sulfide trends before its concentration increases to the point that gaseous H_2S is released from the mud. Draeger tubes are used for spot measurement of hydrogen sulfide, as a backup or as a check on other sensing equipment.

References

- Blundell D. (2005). Processes of tectonism, magmatism and mineralization: Lessons from Europe. Ore Geology Reviews. 27. P. 340. Doi:10.1016/j.oregeorev.2005.07.003. ISBN 9780444522337

- Petroleum-exploration, upstream: oil-gasportal.com, Retrieved 22 April, 2019

- Bousso, Ron (January 18, 2019). "After billion-barrel bonanza, BP goes global with seismic tech". Www.reuters.com. Retrieved January 18, 2019

- Petroleum-production, technology: britannica.com, Retrieved 23 May, 2019

- F., Worthington, Paul (2010-10-01). "Net Pay--What Is It? What Does It Do? How Do We Quantify It? How Do We Use It?". SPE Reservoir Evaluation & Engineering. 13 (05). Doi:10.2118/123561-PA. ISSN 1094-6470. Archived from the original on 2017-03-12

- Types-oil-and-gas-exploration-methods, topics: profolus.com, Retrieved 24 June, 2019

- Freudenrich, Ph.D, Craig; Strickland, Jonathan. "How Oil Drilling Works". Retrieved 21 September 2013

- Spontaneous-(SP)-log: petrowiki.org, Retrieved 25 July, 2019

- Angelakis, Andreas N.; Mays, Larry W.; Koutsoyiannis, Demetris; Mamassis, Nikos (2012). Evolution of Water Supply Through the Millennia. Iwa Publishing (published January 1, 2012). Pp. 202–203. ISBN 978-1843395409

- Drilling-mud, technology: britannica.com, Retrieved 26 August, 2019

- Kleinberg, Robert L.; Jackson, Jasper A. (2001). "An introduction to the history of NMR well logging". Concepts in Magnetic Resonance. 13 (6): 340–342. Doi:10.1002/cmr.1018

- Mud-logging: petrowiki.org, Retrieved 27 January, 2019

- "How the Gulf Crisis Began and Ended (The Gulf Crisis and Japan's Foreign Policy)". Ministry of Foreign Affairs of Japan. Retrieved 28 January 2014

- Fractional-distillation, encyclopedia: energyeducation.ca, Retrieved 28 February, 2019

4

Petroleum Industry and Related Processes

Petroleum industry deals with the processes of exploration, extraction, refining, transporting and marketing of petroleum products. It is related to various industries such as gas industry, oil shale industry and oil refinery. This chapter discusses in detail the petroleum industry and related processes like the petroleum refining process.

Oil Industry, also known as the Petroleum Industry, includes exploration, production, Refining, transportation of Crude and Petroleum Products via rail, road, pipelines or sea (Oil Tankers) and marketing of petroleum products. It plays very vital role in the global economy of any nation because it promotes industrialization. It is divided into three major components, i.e., upstream, midstream and downstream.

The Oil Industry is broadly classified into three sectors:

- Upstream,
- Midstream,
- Downstream.

The upstream sector includes well completions and drilling operations. Exploration and production organizations perform 2D and 3D seismic surveys of the ground to search for hydrocarbon formations of crude oil and natural gas. Once such formations are discovered or explored, exploratory wells are drilled and subsequently drilling is carried out. Some of the upstream sector organizations are: Chevron, BP, Shell, Exxon-Mobil, ConocoPhillips, Total and many more.

The midstream sector involves the transportation of oil or refined petroleum products to various refineries or industries for further processing or usage mainly through rail, tankers and pipelines.

The Downstream sector includes refineries that process crude oil into petroleum products such as LPG, Naphtha, Motor Spirit, High Speed Diesel, Jet Kerosene, Furnace Oil, and heavy products. Once these products are produced, refineries marketing agencies or Oil Marketing Companies (OMC's) sell these petroleum products to various geographical locations.

Working of Oil and Gas Industry

The oil and gas industry is loaded with abstruse terms that can overwhelm an investor new to the sector. This introduction to the industry and its key concepts and measurements can help anyone understand the fundamentals of the companies involved.

Crude oil and natural gas are naturally occurring substances that are found in rock in the Earth's crust. Oil and gas are organic materials that are created by the compression of the remains of plants and animals in sedimentary rock such as sandstone, limestone, and shale.

The sedimentary rock itself is a product of deposits in ancient oceans and other bodies of water. As layers of sediment were deposited on the ocean floor, the decaying remains of plants and animals were integrated into the forming rock. The organic material eventually transforms into oil and gas after being exposed to specific temperatures and pressure ranges deep within the Earth's crust.

Oil and gas are less dense than water, so they migrate through porous sedimentary source rock toward the Earth's surface. When the hydrocarbons are trapped beneath less-porous cap rock, an oil and gas reservoir is formed. These reservoirs of oil and gas are our sources for crude oil and gas.

Hydrocarbons are brought to the surface by drilling through the cap rock and into the reservoir. Once the drill bit reaches the reservoir, a productive oil or gas well can be constructed and the hydrocarbons can be pumped to the surface.

When the drilling activity does not find commercially viable quantities of hydrocarbons, the well is classified as a dry hole. These are typically plugged and abandoned.

Exploration and Production Companies

Exploration and production (E&P) companies find hydrocarbon reservoirs, drill oil and gas wells, extract these raw materials, and sell them to be refined by other companies into products such as gasoline.

This activity is often referred to as upstream oil and gas activity. Today, hundreds of public E&P companies are listed on U.S. stock exchanges. Virtually all cash flow and income statement line items of E&P companies are directly related to oil and gas production.

Understanding Oil Production Numbers

E&P companies measure oil production in barrels. One barrel, usually abbreviated as bbl, is 42 U.S. gallons. Companies often describe production in terms of bbl per day or bbl per quarter.

A common methodology in the oil patch is to use a prefix of "m" to indicate 1,000 and a prefix of "mm" to indicate one million. Therefore, 1,000 barrels is commonly denoted as mbbl and one million barrels is denoted as mmbbl. For example, when an E&P company reports production of seven mbbl per day, it means 7,000 barrels of oil per day.

Gas Production Numbers

Gas production is described in terms of standard cubic feet, which is a measure of the quantity of gas at 60 degrees Fahrenheit and 14.65 pounds per square inch of pressure. Similar to the convention for oil, the term mmcf means one million cubic feet of gas. One billion cubic feet is denoted as Bcf, and one trillion cubic feet is denoted as Tcf.

Note that gas market prices are sold on the New York Mercantile Exchange futures market in quantities based on one million British thermal units, or mmbtu, which is roughly equivalent to 970 cubic feet of gas. Investors frequently think of an mcf of gas as being equivalent to one mmbtu.

E&P companies often describe their production in units of barrels of oil equivalent (BOE). To calculate BOE, companies usually convert gas production into oil equivalent production. In this calculation, one BOE has the energy equivalent of 6,040 cubic feet of gas or roughly one bbl to 6 mcf.

Oil quantity can be converted into gas quantity in a similar fashion, and gas producers often refer to production in terms of gas equivalency using the term mcfe.

Note that the energy conversion basis often is not reflected in the respective market prices of oil and gas.

E&P companies report their oil and gas reserves—the quantity of oil and gas they own that is still in the ground—in the same bbl and mcf terms. Reserves are often used to value E&P companies and make predictions for their revenue and carnings.

The value of reserves is not a GAAP figure and is not directly booked into a company's financial statements.

New reserves, of course, are the primary source of future revenue, so E&P companies spend a lot of time and money exploring for new petroleum reserves. If an E&P

company stops exploring, it will generate revenue from a finite and depleting quantity of petroleum and revenue inevitably will decline over time.

E&P companies can only maintain or grow a revenue base by acquiring or finding new reserves.

Drilling and Service Companies

E&P companies do not usually own their own drilling equipment or employ a drilling rig staff. Instead, they hire contract drilling companies to drill wells for them.

Contract drilling companies generally charge for their services based on the amount of time they work for an E&P company. They do not generate revenue that is tied directly to oil and gas production as is the case for E&P companies.

Once a well is drilled, various activities are involved in generating and maintaining its production over time. These include well logging, cementing, casing, perforating, fracturing, and maintenance and are collectively referred to as well servicing.

As is the case for drilling, many public companies are involved with well service activity. The revenue of service companies is tied to the activity level in the oil and gas industry, sometimes measured by the rig count. That is the number of rigs working in the United States at any given time.

Oil Refinery

An oil refinery or petroleum refinery is an industrial process plant where crude oil is transformed and refined into more useful products such as petroleum naphtha, gasoline, diesel fuel, asphalt base, heating oil, kerosene, liquefied petroleum gas, jet fuel and fuel oils. Petrochemicals feed stock like ethylene and propylene can also be produced directly by cracking crude oil without the need of using refined products of crude oil such as naphtha.

Oil refineries are typically large, sprawling industrial complexes with extensive piping running throughout, carrying streams of fluids between large chemical processing units, such as distillation columns. In many ways, oil refineries use much of the technology, and can be thought of, as types of chemical plants.

The crude oil feedstock has typically been processed by an oil production plant. There is usually an oil depot at or near an oil refinery for the storage of incoming crude oil feedstock as well as bulk liquid products.

Petroleum refineries are very large industrial complexes that involve many different processing units and auxiliary facilities such as utility units and storage tanks.

Each refinery has its own unique arrangement and combination of refining processes largely determined by the refinery location, desired products and economic considerations.

An oil refinery is considered an essential part of the downstream side of the petroleum industry.

Some modern petroleum refineries process as much as 800,000 to 900,000 barrels (127,000 to 143,000 cubic meters) of crude oil per day.

According to the Oil and Gas Journal in the world a total of 636 refineries were operated on the 31 December 2014 for a total capacity of 87.75 million barrels (13,951,000 m³).

Anacortes Refinery (Marathon), on the north end of March Point southeast of Anacortes, Washington, United States.

A petrochemical refinery in Grangemouth, Scotland.

Operation

Raw or unprocessed crude oil is not generally useful in industrial applications, although "light, sweet" (low viscosity, low sulfur) crude oil has been used directly as a burner fuel to produce steam for the propulsion of seagoing vessels. The lighter elements, however, form explosive vapors in the fuel tanks and are therefore hazardous, especially in warships. Instead, the hundreds of different hydrocarbon molecules in crude oil are separated in a refinery into components that can be used as fuels, lubricants, and feedstocks in petrochemical processes that manufacture such products as plastics, detergents, solvents, elastomers, and fibers such as nylon and polyesters.

Petroleum fossil fuels are burned in internal combustion engines to provide power for ships, automobiles, aircraft engines, lawn mowers, dirt bikes, and other machines. Different boiling points allow the hydrocarbons to be separated by distillation. Since the lighter liquid products are in great demand for use in internal combustion engines, a modern refinery will convert heavy hydrocarbons and lighter gaseous elements into these higher value products.

The oil refinery in Haifa, Israel is capable of processing about 9 million tons
(66 million barrels) of crude oil a year. Its two cooling towers are landmarks of the city's skyline.

Oil can be used in a variety of ways because it contains hydrocarbons of varying molecular masses, forms and lengths such as paraffins, aromatics, naphthenes (or cycloalkanes), alkenes, dienes, and alkynes. While the molecules in crude oil include different atoms such as sulfur and nitrogen, the hydrocarbons are the most common form of molecules, which are molecules of varying lengths and complexity made of hydrogen and carbon atoms, and a small number of oxygen atoms. The differences in the structure of these molecules account for their varying physical and chemical properties, and it is this variety that makes crude oil useful in a broad range of several applications.

Once separated and purified of any contaminants and impurities, the fuel or lubricant can be sold without further processing. Smaller molecules such as isobutane and propylene or butylenes can be recombined to meet specific octane requirements by processes such as alkylation, or more commonly, dimerization. The octane grade of gasoline can also be improved by catalytic reforming, which involves removing hydrogen from hydrocarbons producing compounds with higher octane ratings such as aromatics. Intermediate products such as gasoils can even be reprocessed to break a heavy, long-chained oil into a lighter short-chained one, by various forms of cracking such as fluid catalytic cracking, thermal cracking, and hydrocracking. The final step in gasoline production is the blending of fuels with different octane ratings, vapor pressures, and other properties to meet product specifications. Another method for reprocessing and upgrading these intermediate products (residual oils) uses a devolatilization process to separate usable oil from the waste asphaltene material.

Oil refineries are large scale plants, processing about a hundred thousand to several hundred thousand barrels of crude oil a day. Because of the high capacity, many of the units operate continuously, as opposed to processing in batches, at steady state or nearly steady state for months to years. The high capacity also makes process optimization and advanced process control very desirable.

Major Products

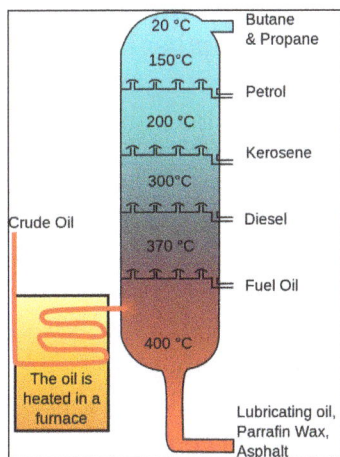

Crude oil is separated into fractions by fractional distillation. The fractions at the top of the fractionating column have lower boiling points than the fractions at the bottom. The heavy bottom fractions are often cracked into lighter, more useful products. All of the fractions are processed further in other refining units.

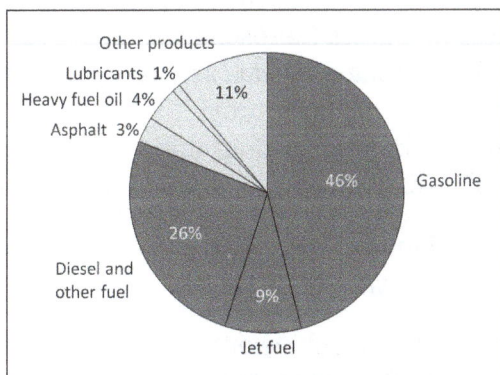

A breakdown of the products made from a typical barrel of US oil.

Petroleum products are materials derived from crude oil (petroleum) as it is processed in oil refineries. The majority of petroleum is converted to petroleum products, which includes several classes of fuels.

Oil refineries also produce various intermediate products such as hydrogen, light hydrocarbons, reformate and pyrolysis gasoline. These are not usually transported but instead are blended or processed further on-site. Chemical plants are thus often adjacent to oil refineries or a number of further chemical processes are integrated into it. For example, light hydrocarbons are steam-cracked in an ethylene plant, and the produced ethylene is polymerized to produce polyethene.

Because technical reasons and environment protection demand a very low sulfur content in all but the heaviest products, it is transformed to hydrogen sulfide via catalytic hydrodesulfurization and removed from the product stream via amine gas treating. Using the Claus process, hydrogen sulfide is afterwards transformed to elementary sulfur to be sold to the chemical industry. The rather large heat energy freed by this process is directly used in the other parts of the refinery. Often an electrical power plant is combined into the whole refinery process to take up the excess heat.

According to the composition of the crude oil and depending on the demands of the market, refineries can produce different shares of petroleum products. The largest share of oil products is used as "energy carriers", i.e. various grades of fuel oil and gasoline. These fuels include or can be blended to give gasoline, jet fuel, diesel fuel, heating oil, and heavier fuel oils. Heavier (less volatile) fractions can also be used to produce asphalt, tar, paraffin wax, lubricating and other heavy oils. Refineries also produce other chemicals, some of which are used in chemical processes to produce plastics and other useful materials. Since petroleum often contains a few percent sulfur-containing molecules, elemental sulfur is also often produced as a petroleum product. Carbon, in the form of petroleum coke, and hydrogen may also be produced as petroleum products. The hydrogen produced is often used as an intermediate product for other oil refinery processes such as hydrocracking and hydrodesulfurization.

Petroleum products are usually grouped into four categories: light distillates (LPG, gasoline, naphtha), middle distillates (kerosene, jet fuel, diesel), heavy distillates and residuum (heavy fuel oil, lubricating oils, wax, asphalt). These require blending various feedstocks, mixing appropriate additives, providing short term storage, and preparation for bulk loading to trucks, barges, product ships, and railcars. This classification is based on the way crude oil is distilled and separated into fractions.

- Gaseous fuel such as Liquified petroleum gas and propane, stored and shipped in liquid form under pressure.

- Lubricants (produces light machine oils, motor oils, and greases, adding viscosity stabilizers as required), usually shipped in bulk to an offsite packaging plant.

- Paraffin wax, used in the packaging of frozen foods, among others. May be shipped in bulk to a site to prepare as packaged blocks. Used for wax emulsions, construction board, matches, candles, rust protection, and vapor barriers.

- Sulfur (or sulfuric acid), byproducts of sulfur removal from petroleum which may have up to a couple percent sulfur as organic sulfur-containing compounds. Sulfur and sulfuric acid are useful industrial materials. Sulfuric acid is usually prepared and shipped as the acid precursor oleum.

- Bulk tar shipping for offsite unit packaging for use in tar-and-gravel roofing.

- Asphalt used as a binder for gravel to form asphalt concrete, which is used for paving roads, lots, etc. An asphalt unit prepares bulk asphalt for shipment.

- Petroleum coke, used in specialty carbon products like electrodes or as solid fuel.

- Petrochemicals are organic compounds that are the ingredients for the chemical industry, ranging from polymers and pharmaceuticals, including ethylene and benzene-toluene-xylenes ("BTX") which are often sent to petrochemical plants for further processing in a variety of ways. The petrochemicals may be olefins or their precursors, or various types of aromatic petrochemicals.

- Gasoline.

- Naphtha.

- Kerosene and related jet aircraft fuels.

- Diesel fuel and Fuel oils.

- Heat.

- Electricity.

Over 6,000 items are made from petroleum waste by-products including: fertilizer, floor coverings, perfume, insecticide, petroleum jelly, soap, vitamin capsules.

| Sample of Crude oil (petroleum) | Cylinders of Liquified petroleum gas | Sample of Gasoline | Sample of Kerosene |

| Sample of Diesel fuel | Motor oil | Pile of asphalt-covered | Sulphur |

Chemical Processes Found in a Refinery

- Desalter unit washes out salt from the crude oil before it enters the atmospheric distillation unit.

- Crude Oil Distillation unit (Atmospheric distillation): Distills the incoming crude oil into various fractions for further processing in other units.

- Vacuum distillation further distills the residue oil from the bottom of the crude oil distillation unit. The vacuum distillation is performed at a pressure well below atmospheric pressure.

Storage tanks and towers at Shell Puget Sound Refinery
(Shell Oil Company), Anacortes, Washington.

- Naphtha hydrotreater unit uses hydrogen to desulfurize naphtha from atmospheric distillation. Must hydrotreat the naphtha before sending to a catalytic reformer unit.

- Catalytic reformer converts the desulfurized naphtha molecules into higher-octane molecules to produce reformate (reformer product). The reformate has higher content of aromatics and cyclic hydrocarbons which is a component of the end-product gasoline or petrol. An important byproduct of a reformer is hydrogen released during the catalyst reaction. The hydrogen is used either in the hydrotreaters or the hydrocracker.

- Distillate hydrotreater desulfurizes distillates (such as diesel) after atmospheric distillation. Uses hydrogen to desulfurize the naphtha fraction from the crude oil distillation or other units within the refinery.

- Fluid Catalytic Cracker (FCC) upgrades the heavier, higher-boiling fractions from the crude oil distillation by converting them into lighter and lower boiling, more valuable products.

- Hydrocracker uses hydrogen to upgrade heavy residual oils from the vacuum distillation unit by thermally cracking them into lighter, more valuable reduced viscosity products.

- Merox desulfurize LPG, kerosene or jet fuel by oxidizing mercaptans to organic disulfides.

- Alternative processes for removing mercaptans are known, e.g. doctor sweetening process and caustic washing.

- Coking units (delayed coking, fluid coker, and flexicoker) process very heavy residual oils into gasoline and diesel fuel, leaving petroleum coke as a residual product.

- Alkylation unit uses sulfuric acid or hydrofluoric acid to produce high-octane components for gasoline blending. Converts isobutane and butylenes into *alkylate*, which is a very high-octane component of the end-product gasoline or petrol.

- Dimerization unit converts olefins into higher-octane gasoline blending components. For example, butenes can be dimerized into isooctene which may subsequently be hydrogenated to form isooctane. There are also other uses for dimerization. Gasoline produced through dimerization is highly unsaturated and very reactive. It tends spontaneously to form gums. For this reason the effluent from the dimerization need to be blended into the finished gasoline pool immediately or hydrogenated.

- Isomerization converts linear molecules such as normal pentane to higher-octane branched molecules for blending into gasoline or feed to alkylation units. Also used to convert linear normal butane into isobutane for use in the alkylation unit.

- Steam reforming converts natural gas into hydrogen for the hydrotreaters and the hydrocracker.

- Liquified gas storage vessels store propane and similar gaseous fuels at pressure sufficient to maintain them in liquid form. These are usually spherical vessels or "bullets" (i.e., horizontal vessels with rounded ends).

- Amine gas treater, Claus unit, and tail gas treatment convert hydrogen sulfide from hydrodesulfurization into elemental sulfur. The large majority of the 64,000,000 metric tons of sulfur produced worldwide in 2005 was byproduct sulfur from petroleum refining and natural gas processing plants.

- Sour water stripper Uses steam to remove hydrogen sulfide gas from various wastewater streams for subsequent conversion into end-product sulfur in the Claus unit.

- Cooling towers circulate cooling water, boiler plants generates steam for steam generators, and instrument air systems include pneumatically operated control valves and an electrical substation.

- Wastewater collection and treating systems consist of API separators, dissolved air flotation (DAF) units and further treatment units such as an activated sludge biotreater to make water suitable for reuse or for disposal.

- Solvent refining use solvent such as cresol or furfural to remove unwanted, mainly aromatics from lubricating oil stock or diesel stock.

- Solvent dewaxing remove the heavy waxy constituents petrolatum from vacuum distillation products.

- Liquified gas (LPG) storage vessels for propane and similar gaseous fuels at a pressure sufficient to maintain them in liquid form. These are usually spherical vessels or *bullets* (horizontal vessels with rounded ends).

- Storage tanks for storing crude oil and finished products, usually vertical, cylindrical vessels with some sort of vapour emission control and surrounded by an earthen berm to contain spills.

Flow Diagram of Typical Refinery

The image below is a schematic flow diagram of a typical oil refinery that depicts the various unit processes and the flow of intermediate product streams that occurs between the inlet crude oil feedstock and the final end products. The diagram depicts only one of the literally hundreds of different oil refinery configurations. The diagram also does not include any of the usual refinery facilities providing utilities such as steam, cooling water, and electric power as well as storage tanks for crude oil feedstock and for intermediate products and end products.

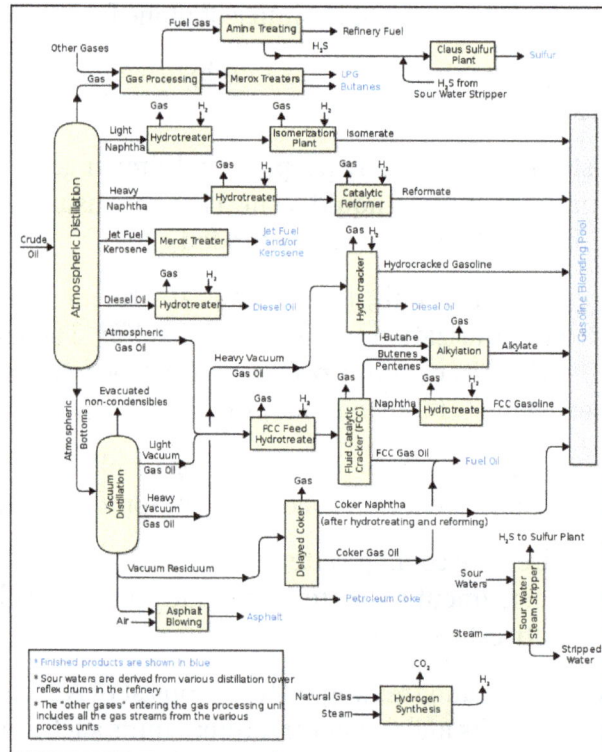

Schematic flow diagram of a typical oil refinery.

There are many process configurations other than that depicted above. For example, the vacuum distillation unit may also produce fractions that can be refined into end products such as: spindle oil used in the textile industry, light machinery oil, motor oil, and various waxes.

The Crude Oil Distillation Unit

The crude oil distillation unit (CDU) is the first processing unit in virtually all petroleum refineries. The CDU distills the incoming crude oil into various fractions of different boiling ranges, each of which are then processed further in the other refinery processing units. The CDU is often referred to as the *atmospheric distillation unit* because it operates at slightly above atmospheric pressure.

Below is a schematic flow diagram of a typical crude oil distillation unit. The incoming crude oil is preheated by exchanging heat with some of the hot, distilled fractions and other streams. It is then desalted to remove inorganic salts (primarily sodium chloride).

Following the desalter, the crude oil is further heated by exchanging heat with some of the hot, distilled fractions and other streams. It is then heated in a fuel-fired furnace (fired heater) to a temperature of about 398 °C and routed into the bottom of the distillation unit.

The cooling and condensing of the distillation tower overhead is provided partially by exchanging heat with the incoming crude oil and partially by either an air-cooled or water-cooled condenser. Additional heat is removed from the distillation column by a pumparound system as shown in the diagram below.

As shown in the flow diagram, the overhead distillate fraction from the distillation column is naphtha. The fractions removed from the side of the distillation column at various points between the column top and bottom are called *sidecuts*. Each of the sidecuts (i.e., the kerosene, light gas oil and heavy gas oil) is cooled by exchanging heat with the incoming crude oil. All of the fractions (i.e., the overhead naphtha, the sidecuts and the bottom residue) are sent to intermediate storage tanks before being processed further.

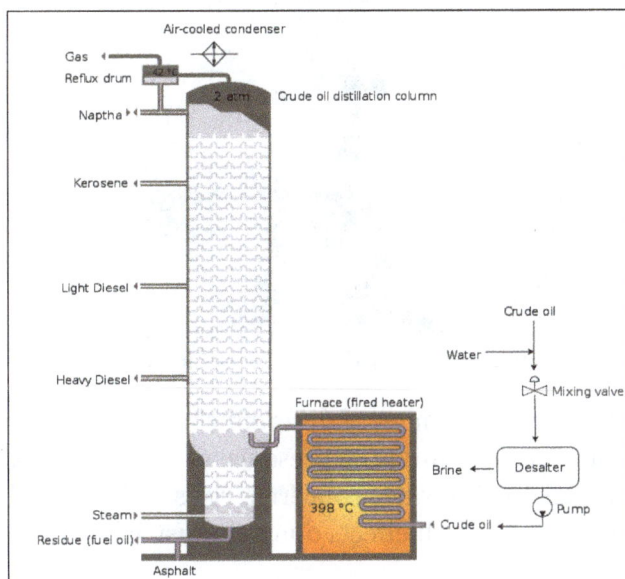

Schematic flow diagram of a typical crude oil distillation unit as used in petroleum crude oil refineries.

Location of Petroleum Refineries

A party searching for a site to construct a refinery or a chemical plant needs to consider the following issues:

- The site has to be reasonably far from residential areas.

- Infrastructure should be available for supply of raw materials and shipment of products to markets.

- Energy to operate the plant should be available.

- Facilities should be available for waste disposal.

Refineries which use a large amount of steam and cooling water need to have an abundant source of water. Oil refineries therefore are often located nearby navigable rivers or on a sea shore, nearby a port. Such location also gives access to transportation by river or by sea. The advantages of transporting crude oil by pipeline are evident, and oil companies often transport a large volume of fuel to distribution terminals by pipeline. Pipeline may not be practical for products with small output, and rail cars, road tankers, and barges are used.

Petrochemical plants and solvent manufacturing (fine fractionating) plants need spaces for further processing of a large volume of refinery products for further processing, or to mix chemical additives with a product at source rather than at blending terminals.

Safety and Environment

Fire-extinguishing operations after the Texas City Refinery explosion.

The refining process releases a number of different chemicals into the atmosphere and a notable odor normally accompanies the presence of a refinery. Aside from air pollution impacts there are also wastewater concerns, risks of industrial accidents such as fire and explosion, and noise health effects due to industrial noise.

Many governments worldwide have mandated restrictions on contaminants that refineries release, and most refineries have installed the equipment needed to comply with the requirements of the pertinent environmental protection regulatory agencies. In the United States, there is strong pressure to prevent the development of new refineries, and no major refinery has been built in the country since Marathon's Garyville, Louisiana facility in 1976. However, many existing refineries have been expanded during that time. Environmental restrictions and pressure to prevent construction of new refineries may have also contributed to rising fuel prices in the United States. Additionally, many refineries (more than 100 since the 1980s) have closed due to obsolescence and merger activity within the industry itself.

Environmental and safety concerns mean that oil refineries are sometimes located some distance away from major urban areas. Nevertheless, there are many instances where refinery operations are close to populated areas and pose health risks. In California's Contra Costa County and Solano County, a shoreline necklace of refineries, built in the early 20th century before this area was populated, and associated chemical plants are adjacent to urban areas in Richmond, Martinez, Pacheco, Concord, Pittsburg, Vallejo and Benicia, with occasional accidental events that require "shelter in place" orders to the adjacent populations. A number of refineries are located in Sherwood Park, Alberta, directly adjacent to the City of Edmonton. The Edmonton metro area has a population of over 1,000,000 residents.

NIOSH criteria for occupational exposure to refined petroleum solvents have been available since 1977.

Corrosion

Refinery of Slovnaft in Bratislava.

Oil refinery in Iran.

Corrosion of metallic components is a major factor of inefficiency in the refining process. Because it leads to equipment failure, it is a primary driver for the refinery maintenance schedule. Corrosion-related direct costs in the U.S. petroleum industry as of 1996 were estimated at US $3.7 billion.

Corrosion occurs in various forms in the refining process, such as pitting corrosion from water droplets, embrittlement from hydrogen, and stress corrosion cracking from sulfide attack. From a materials standpoint, carbon steel is used for upwards of 80 per cent of refinery components, which is beneficial due to its low cost. Carbon steel is resistant to the most common forms of corrosion, particularly from hydrocarbon impurities at temperatures below 205 °C, but other corrosive chemicals and environments prevent its use everywhere. Common replacement materials are low alloy steels containing chromium and molybdenum, with stainless steels containing more chromium dealing with more corrosive environments. More expensive materials commonly used are nickel, titanium, and copper alloys. These are primarily saved for the most problematic areas where extremely high temperatures and very corrosive chemicals are present.

Corrosion is fought by a complex system of monitoring, preventative repairs and careful use of materials. Monitoring methods include both offline checks taken during maintenance and online monitoring. Offline checks measure corrosion after it has occurred, telling the engineer when equipment must be replaced based on the historical information they have collected. This is referred to as preventative management.

Online systems are a more modern development, and are revolutionizing the way corrosion is approached. There are several types of online corrosion monitoring technologies such as linear polarization resistance, electrochemical noise and electrical resistance. Online monitoring has generally had slow reporting rates in the past (minutes or hours) and been limited by process conditions and sources of error but newer technologies can report rates up to twice per minute with much higher accuracy (referred to as real-time monitoring). This allows process engineers to treat corrosion as another process variable that can be optimized in the system. Immediate responses to process changes allow the control of corrosion mechanisms, so they can be minimized while also maximizing production output. In an ideal situation having online corrosion information that is accurate and real-time will allow conditions that cause high corrosion rates to be identified and reduced. This is known as predictive management.

Materials methods include selecting the proper material for the application. In areas of minimal corrosion, cheap materials are preferable, but when bad corrosion can occur, more expensive but longer lasting materials should be used. Other materials methods come in the form of protective barriers between corrosive substances and the equipment metals. These can be either a lining of refractory material such as standard Portland cement or other special acid-resistant cements that are shot onto the inner surface of the vessel. Also available are thin overlays of more expensive metals that protect cheaper metal against corrosion without requiring lots of material.

Instrumentation in Petrochemical Industries

Instrumentation is used to monitor and control the process plant in the oil, gas and petrochemical industries. Instrumentation comprises sensor elements, signal transmitters, controllers, indicators and alarms, actuated valves, logic circuits and operator interfaces.

The Elements of Instrumentation

Instrumentation includes sensing devices to measure process parameters such as pressure, temperature, liquid level, flow, velocity, composition, density, weight; and mechanical and electrical parameters such as vibration, position, current and voltage.

The measured value of a parameter can be displayed and recorded either locally and in a control room. If the measured variable exceeds pre-defined limits an alarm may be provided to warn the operating personnel of a potential problem. Automatic executive action can also be taken by the instrumentation to close or open shutdown valves and dampers, or to trip (stop) pumps and compressors.

Correct operation of the petrochemical process plant is achieved through the action of control loops. These automatically maintain and control the pressure, temperature, liquid level and flowrate of fluid in vessels and piping. Such control loops generally work by comparing the measured value of a parameter on the plant, eg. pressure, with a pre-determined set point. Any difference between the measured variable and the set point generates a signal which is used to modulate the position of a control valve (the final element) to maintain the measured variable at the set point.

Valves can be actuated by an electric motor, hydraulic fluid or air. For air-operated control valves, electrical signals from the control system are converted to an air pressure for the valve actuator in a current/pneumatic I/P converter. Upon loss of pneumatic or hydraulic pressure valves can be configured to fail to an open (FO) or fail to a closed (FC) position.

Some instrumentation is self actuating. For example, pressure regulators maintain a constant pre-set pressure, and rupture discs and pressure safety valves open at pre-set pressures.

Instrumentation includes facilities for operating personnel to intervene in the plant either locally or from a control room. Personnel can open or close valves, change set points, start and stop pumps or compressors, over-ride shutdown functions (in specific controlled circumstances such as during start-up).

Temperature Instrumentation

Measurement of temperature of fluids in the petrochemical industry is undertaken by temperature elements (TE). These can be Thermocouples or Platinum Resistance Temperature Detectors (RTD's). The latter are used for their good temperature response. Local temperature indicators (TI) are located on the inlet and outlet streams of heat exchangers to monitor the performance of the heat exchanger.

In industrial applications gaseous or liquid fluids may need to be heated or cooled. This duty is undertaken in a heat exchanger, whereby the fluid is heated or cooled by heat transfer with a second fluid such as water, glycol, hot oil or another process fluid (the heating or cooling medium). Temperature control is used to maintain the desired temperature of the first fluid. A temperature sensor transmitter (TT) is located in the first fluid at its outlet from the heat exchanger. This measured temperature is fed into the temperature controller (TIC) where it is compared to the desired set point temperature. The output of the controller, which is related to the difference between the measured variable and the set point, is fed to a control valve (TCV) in the second fluid to adjust the flow of the heating or cooling medium. In the case of a fluid being cooled, if the temperature of the fluid rises the temperature controller acts to open the TCV increasing the flow of the cooling medium which increases the heat transfer and reduces the temperature of the first fluid. Conversely if the temperature falls the controller acts to close the TCV which reduces the heat transfer increasing the temperature of the first fluid. In the case of heating medium with the falling temperature of the first fluid the controller would act to open the TCV to increase the flow of heating medium thereby raising the temperature of the first fluid. The controller (TIC) may also generate high (TAH) and low temperature (TAL) alarms to warn operating personnel of a potential problem.

Fin fan coolers use air to cool gases and liquids. The temperature of fluid is controlled (TIC) by opening or closing dampers on the cooler or adjusting the speed of the fan or the pitch angle of the fan blades thereby increasing or decreasing the flow of air.

Temperature monitoring and control instrumentation is used in fired heaters and furnaces to adjust the fuel flow valve (FCV) to maintain a desired thermal output. Waste heat recovery units (WHRU) are used to extract heat from the flow of hot exhaust gases from a gas turbine to heat a fluid (heating medium). Instrumentation includes controllers to maintain a desired temperature of the heating medium by closing or opening dampers in the exhaust gas flow.

Low temperature alarms (TSL) are used where cold fluids could be routed to pipework which is not suitable for cold service. Instrumentation may include an initial alarm (TAL) and then a shutdown action (TSLL) to close a shutdown valve (XV).

Temperature sensors (TE) are used to indicate that plant flares have been unintentionally extinguished (BAL), perhaps due to insufficient flowrate of gases to maintain a flame.

Pressure Instrumentation

Many oil, gas and petrochemical processes are undertaken at specific pressures. Pressure is measured by pressure sensors (PE) which are arranged to transmit pressure (PT) signals to pressure controllers (PIC). Pressure vessels and tanks are also usually provided with local pressure indicators (PI).

Pressure in the petrochemical industry is frequently controlled by maintaining a constant pressure in the upper gas space of a vessel. The controller (PIC) adjusts the setting on a pressure control valve (PCV) that feeds gas forward to the next stage of the process. A rising pressure in the vessel results in the PCV opening to feed more gas forward. If the pressure continues to rise some controllers then act to open a second PCV that feeds excess gas to the flare system. The pressure transmitter is configured to provide warning alarms (PAL and PAH) if the pressure exceeds set high and low limits. If these limits are further exceeded (PALL and PAHH) an automatic shutdown of the system is initiated which includes closure of the inlet valves of the vessel. The pressure sensor (PT) that initiates a shutdown is a separate instrument loop from the PT associated with the pressure control loop to mitigate common mode failures and to ensure greater reliability of the shutdown function.

The operation of hydrocyclones is controlled by pressure instrumentation that maintains fixed differential pressures between the inlet and the oil and water outlets.

Turbo-expanders are controlled by maintaining the inlet pressure (PIC) at a constant value by controlling the angle of the expander inlet vanes. A split range pressure controller may also modulate a Joule-Thomson valve across the turbo-expander.

Pressure in blanketed tanks is maintained by self actuating pressure control valves (PCVs). As liquid is withdrawn from the tank the pressure in the gas space falls. The blanket gas supply valve opens to maintain the pressure. As the tank fills with liquid the pressure rises and a vent gas valve open to vent gas to atmosphere or a vent system.

Two important items of pressure instrumentation are rupture (bursting) discs (PSE) and pressure relief or pressure safety valves (PSV). Both are self-actuating and are designed to open at a preset pressure to provide an essential safety function on the petrochemical plant.

Flow Instrumentation

The throughput of a petrochemical plant is measured and controlled by flow instrumentation.

Flow measuring devices devices (FE) include vortex, positive displacement (PD), differential pressure (DP), coriolis, ultrasonic, and rotameters.

The flow through compressors is, at its simplest form, controlled by measuring the flow (FT) through the machine at the discharge and controlling the speed (FIC/SIC) of the prime mover (electric motor or gas turbine) that is driving the compressor. Anti-surge control ensures a minimum flow of fluid through the compressor. This requires measurement of flow (FT) at the discharge plus measurements of the suction and discharge pressures (PT) and temperatures (TT) of the fluid flowing through the compressor. The anti-surge controller (FIC) modulates a control valve (FCV) which recycles cooled gas from downstream of the compressor after-cooler back to the suction of the compressor. Low flow alarms (FAL) provide a warning indication to operating personnel.

Large process pumps are provided with minimum flow protection. This comprises measurement of flow (FT) at the pump discharge, this measurement is an input to a flow controller (FIC) whose set point is the minimum flow required through the pump.

As the flow reduces to the minimum flow value the controller acts to open a flow control valve (FCV) to recycle fluid from the discharge back to the suction of the pump.

Flow metering (FIQ) is required where custody transfer of fluids takes place, such as an outgoing pipeline or at a tanker loading station. This requires accurate measurement of the flow with inputs such as liquid density.

Flare and vent systems need to be purged to prevent air ingress and the formation of potentially explosive mixtures. The flowrate of purge gas is set by rotameter (FIC) or fixed orifice plate (FO). A low flow alarm (FAL) provides a warning indication to operating personnel that the purge flow has reduced significantly.

Pipelines are monitored by measuring the flowrate of fluid at each end, a discrepancy (FDA) may indicate a leak in the pipeline.

Level Instrumentation

The level measurement of liquids in pressure vessels and tanks in the petrochemical industry is undertaken by differential pressure level meters, radar, magnetostrictive, nucleonic, magnetic float and pneumatic bubbler instruments.

Level instrumentation determines the height of liquids by measuring the position of a gas/liquid or liquid/liquid interface within the vessel or tank. Such interfaces include oil/gas, oil/water, condensate/water, glycol/condensate, etc. Local indication (LI) includes sight glasses which show the liquid level directly through a vertical glass tube attached to the vessel/tank.

Phase interfaces are maintained at a constant level by level transmitters (LT) transmitting a signal to a level controller (LIC) which compares the measured value with the desired set point. The difference is sent as a signal to a level control valve (LCV) on the liquid outlet from the vessel. As the level rises the controller acts to open the valve to draw off liquid to reduce the level. Similarly as the levels fall the controller acts to close the LCV to reduce outflow of fluid.

For some vessels liquid is pumped out. The controller (LIC) acts to start and stop the pump within a specified band. For example, start the pump when the level rises to 0.6m, stop the pump when the level falls to 0.4m.

High and low level alarms (LAH and LAL) warm operating personnel that levels have gone outside predefined limits. Further deviation (LAHH and LALL) initiates a shutdown either to close emergency shutdown valves (ESDV) on the inlet to the vessel or on the liquid outlet lines. As with high and low pressure instrumentation the shutdown function should comprise an independent measurement loop to prevent a common mode failure. Loss of liquid level in the vessel may lead to gas blowby where high pressure gas flows to the downstream vessel through the liquid outlet line. The integrity of the downstream vessel may thereby be compromised. In addition high liquid level in the vessel may lead to carryover of liquid into the gas outlet which could damage down-stream equipment such as gas compressors.

High liquid level in a flare drum may lead to undesirable carryover of liquid to the flare. A high-high liquid level (LSHH) in the flare drum may initiate a plant shutdown.

One of the problems with a significant number of technologies is that they are installed through a nozzle and are exposed to products. This can create several problems, especially when retrofitting new equipment to vessels that have already been stress relieved, as it may not be possible to fit the instrument at the location required. Also, as the measuring element is exposed to the contents within the vessel, it may either attack or coat the instrument causing it to fail in service. One of the most reliable methods for measuring level is using a Nuclear gauge, as it is installed outside the vessel and doesn't normally require a nozzle for bulk level measurement. The measuring element is installed outside the process and can be maintained in normal operation without taking a shutdown. Shutdown is only required for an accurate calibration.

Analyser Instrumentation

A wide range of analysis instruments are used in the oil, gas and petrochemical industries.

- Chromatography – to measure the quality of product or reactants.

- Density (oil) – for custody metering of liquids.

- Dewpoint (water dewpoint and hydrocarbon dewpoint) to check the efficiency of dehydration or dewpoint control plant.

- Electrical conductivity – to measure the effectiveness of potable water reverse osmosis plant.

- Oil-in-water – prior to discharge of water into the environment.

- pH of reactants and products.

- Sulphur content – to check the efficiency of gas sweetening plant.

Most instruments function continuously and provide a log of data and trends. Some analyser instruments are configured to alarm (AAH) if a measurement reaches a critical level.

Other Instrumentation

Major pumps and compressors may be provided with vibration sensors (VT) to provide operating personnel with a warning (VA) of potential mechanical problems with the machine.

Rupture discs (PSE) and Pressure safety valves (PSV) are self-actuated and provide no immediate indication that they have ruptured or lifted. Instrumentation such as pressure alarms (PXA) or movement alarms (PZA) may be fitted to indicate that they have operated.

Corrosion coupons and corrosion probes provide a local indication of corrosion rates of fluids flowing in piping.

Pipeline pig launchers and receivers are provided with a pig signaller (XI) to indicate that a pig has been launched or has arrived.

Packaged items of equipment (compressors, diesel engines, electricity generators, etc) will be provided with local vendor supplied instrumentation. If the equipment malfunctions a multivariable signal (UA) may be sent to the control room.

The fire and gas detection system comprises local sensors to detect the presence of gas, smoke or fire. These provide alarms in the control room. Simultaneous detection of multiple sensors initiates action to start firewater pumps and close fire dampers in enclosed spaces.

The petrochemical plant may have several levels of shutdown. A unit shutdown (USD) entails shutdown of one limited unit with the rest of the plant remaining in operation. A production shutdown (PSD) entails shutdown of the entire process plant. An emergency shutdown (ESD) entails complete shutdown of the plant.

Older plant may have local control loops which operate pneumatic (3 – 15 psia) final element actuators. Sensors may also transmit electrical signals (4 – 20mA). Conversion between pneumatic and electrical signals is undertaken by P/I and I/P converters. Control of modern plant is based on a Distributed Control Systems using Fieldbus digital protocols.

Oil Shale Industry

VKG Energia in Estonia.

The oil shale industry is an industry of mining and processing of oil shale—a fine-grained sedimentary rock, containing significant amounts of kerogen (a solid mixture of organic chemical compounds), from which liquid hydrocarbons can be manufactured. The industry has developed in Brazil, China, Estonia and to some extent in Germany and Russia. Several other countries are currently conducting research on their oil shale reserves and production methods to improve efficiency and recovery. Estonia accounted for about 70% of the world's oil shale production in a study published in 2005.

Oil shale has been used for industrial purposes since the early 17th century, when it was mined for its minerals. Since the late 19th century, shale oil has also been used for its oil content and as a low grade fuel for power generation. However, barring countries having significant oil shale deposits, its use for power generation is not particularly widespread. Similarly, oil shale is a source for production of synthetic crude oil and it is seen as a solution towards increasing domestic production of oil in countries that are reliant on imports.

Mining

Oil shale is mined either by traditional underground mining or surface mining techniques. There are several mining methods available, but the common aim of all these methods is to fragment the oil shale deposits in order to enable the transport of shale fragments to a power plant or retorting facility. The main methods of surface mining are *open pit mining* and *strip mining*. An important method of sub-surface mining is the *room-and-pillar method*. In this method, the material is extracted across a horizontal plane while leaving "pillars" of untouched material to support the roof. These pillars reduce the likelihood of a collapse. Oil shale can also be obtained as a by-product of coal mining.

The largest oil shale mine in the world is the Estonia Mine, operated by Enefit Kaevandused. In 2005, Estonia mined 14.8 million tonnes of oil shale. During the same period, mining permits were issued for almost 24 million tonnes, with applications being received for mining an additional 26 million tonnes. In 2008, the Estonian Parliament approved the "National Development Plan for the Use of Oil Shale 2008-2015", which limits the annual extraction of oil shale to 20 million tonnes.

Power Generation

Oil-shale-fired Eesti Power Plant in Narva, Estonia.

Oil shale can be used as a fuel in thermal power plants, wherein oil shale is burnt like coal to drive the steam turbines. As of 2012, there are oil shale-fired power plants in Estonia with a generating capacity of 2,967 megawatts (MW), China, and Germany.

Also Israel, Romania and Russia have run oil shale-fired power plants, but have shut them down or switched to other fuels like natural gas. Jordan and Egypt have announced their plans to construct oil shale-fired power plants, while Canada and Turkey plan to burn oil shale at the power plants along with coal.

Thermal power plants which use oil shale as a fuel mostly employ two types of combustion methods. The traditional method is *Pulverized combustion* (PC) which is used in the older units of oil shale-fired power plants in Estonia, while the more advanced method is *Fluidized bed combustion* (FBC), which is used in the Holcim cement factory in Dotternhausen, Germany, and was used in the Mishor Rotem power plant in Israel. The main FBC technologies are *Bubbling fluidized bed combustion* (BFBC) and *Circulating fluidized bed combustion* (CFBC).

There are more than 60 power plants around the world, which are using CFBC technology for combustion of coal and lignite, but only two new units at Narva Power Plants in Estonia, and one at Huadian Power Plant in China use CFBC technology for combustion of oil shale. The most advanced and efficient oil shale combustion technology is *Pressurized fluidized-bed combustion* (PFBC). However, this technology is still premature and is in its nascent stage.

Oil Extraction

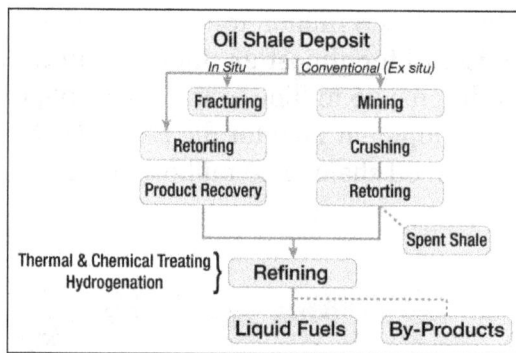

Overview of shale oil extraction.

The major shale oil producers are China and Estonia, with Brazil a distant third, while Australia, USA, Canada and Jordan have planned to set up or restart shale oil production. According to the World Energy Council, in 2008 the total production of shale oil from oil shale was 930,000 tonnes, equal to 17,700 barrels per day (2,810 m³/d), of which China produced 375,000 tonnes, Estonia 355,000 tonnes, and Brazil 200 tonnes. In comparison, production of the conventional oil and natural gas liquids in 2008 amounted 3.95 billion tonnes or 82.12 million barrels per day (13.056×10⁶ m³/d).

Although there are several oil shale retorting technologies, only four technologies are currently in commercial use. These are Kiviter, Galoter, Fushun, and Petrosix. The two main methods of extracting oil from shale are *ex-situ* and *in-situ*. In *ex-situ* method, the oil shale is mined and transported to the retort facility in order to extract the oil. The *in-situ* method converts the kerogen while it is still in the form of an oil shale deposit, and then extracts it via a well, where it rises up as normal petroleum.

Other Industrial uses

Oil shale is used for cement production by Kunda Nordic Cement in Estonia, by Holcim in Germany, and by Fushun cement factory in China. Oil shale can also be used for production of different chemical products, construction materials, and pharmaceutical products, e.g. ammonium bituminosulfonate. However, use of oil shale for production of these products is still very rare and in experimental stages only.

Some oil shales are suitable source for sulfur, ammonia, alumina, soda ash, and nahcolite which occur as shale oil extraction byproducts. Some oil shales can also be used for uranium and other rare chemical element production. During 1946–1952, a marine variety of Dictyonema shale was used for uranium production in Sillamäe, Estonia, and during 1950–1989 alum shale was used in Sweden for the same purpose. Oil shale gas can also be used as a substitute for natural gas. After World War II, Estonian-produced oil shale gas was used in Leningrad and the cities in North Estonia. However, at the current price level of natural gas, this is not economically feasible.

Economics

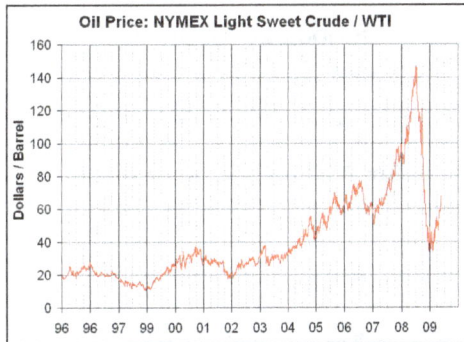

NYMEX light-sweet crude oil prices 1996–2009 (not adjusted for inflation).

The amount of economically recoverable oil shale is unknown. The various attempts to develop oil shale deposits have succeeded only when the cost of shale-oil production in a given region comes in below the price of crude oil or its other substitutes. According to a survey conducted by the RAND Corporation, the cost of producing a barrel of shale oil at a hypothetical surface retorting complex in the United States (comprising a mine, retorting plant, upgrading plant, supporting utilities, and spent shale reclamation), would range between US$70–95 ($440–600/m³), adjusted to 2005 values. Assuming a gradual increase in output after the start of commercial production, the analysis projects a gradual reduction in processing costs to $30–40 per barrel ($190–250/m³) after achieving the milestone of 1 billion barrels (160×10⁶ m³). Royal Dutch Shell has announced that its Shell ICP technology would realize a profit when crude oil prices are higher than $30 per barrel ($190/m³), while some technologies at full-scale production assert profitability at oil prices even lower than $20 per barrel ($130/m³).

To increase the efficiency of oil shale retorting and by this the viability of the shale oil production, researchers have proposed and tested several co-pyrolysis processes, in which other materials such as biomass, peat, waste bitumen, or rubber and plastic wastes are retorted along with the oil shale. Some modified technologies propose combining a fluidized bed retort with a circulated fluidized bed furnace for burning the by-products of pyrolysis (char and oil shale gas) and thereby improving oil yield, increasing throughput, and decreasing retorting time.

In a 1972 publication by the journal *Pétrole Informations*, shale oil production was unfavorably compared to the coal liquefaction. The article stated that coal liquefaction was less expensive, generated more oil, and created fewer environmental impacts than oil shale extraction. It cited a conversion ratio of 650 liters (170 U.S. gal; 140 imp gal) of oil per one tonne of coal, as against 150 liters (40 U.S. gal; 33 imp gal) of shale oil per one tonne of oil shale.

A critical measure of the viability of oil shale as an energy source lies in the ratio of the energy produced by the shale to the energy used in its mining and processing, a ratio known as "Energy Returned on Energy Invested" (EROEI). A 1984 study estimated the EROEI of the various known oil-shale deposits as varying between 0.7–13.3 although known oil-shale extraction development projects assert an EROEI between 3 and 10. According to the World Energy Outlook 2010, the EROEI of *ex-situ* processing is typically 4 to 5 while of *in-situ* processing it may be even as low as 2. However, according to the IEA most of used energy can be provided by burning the spent shale or oil-shale gas.

The water needed in the oil shale retorting process offers an additional economic consideration: this may pose a problem in areas with water scarcity.

Environmental Considerations

Mining oil shale involves a number of environmental impacts, more pronounced in surface mining than in underground mining. These include acid drainage induced by the sudden rapid exposure and subsequent oxidation of formerly buried materials, the introduction of metals including mercury into surface-water and groundwater, increased erosion, sulfur-gas emissions, and air pollution caused by the production of particulates during processing, transport, and support activities. In 2002, about 97% of air pollution, 86% of total waste and 23% of water pollution in Estonia came from the power industry, which uses oil shale as the main resource for its power production.

Oil-shale extraction can damage the biological and recreational value of land and the ecosystem in the mining area. Combustion and thermal processing generate waste material. In addition, the atmospheric emissions from oil shale processing and combustion include carbon dioxide, a greenhouse gas. Environmentalists oppose production and usage of oil shale, as it creates even more greenhouse gases than conventional fossil fuels. Experimental *in situ* conversion processes and carbon capture and storage technologies may reduce some of these concerns in the future, but at the same time they may cause other problems, including groundwater pollution. Among the water contaminants commonly associated with oil shale processing are oxygen and nitrogen heterocyclic hydrocarbons. Commonly detected examples include quinoline derivatives, pyridine, and various alkyl homologues of pyridine (picoline, lutidine).

Water concerns are sensitive issues in arid regions, such as the western US and Israel's Negev Desert, where plans exist to expand oil-shale extraction despite a water shortage. Depending on technology, above-ground retorting uses between one and five barrels of water per barrel of produced shale-oil. A 2008 programmatic environmental impact statement issued by the US Bureau of Land Management stated that surface mining and retort operations produce 2 to 10 U.S. gallons (7.6 to 37.9 l; 1.7 to 8.3 imp gal) of waste water per 1 short ton (0.91 t) of processed oil shale. *In situ* processing, according to one estimate, uses about one-tenth as much water.

Environmental activists, including members of Greenpeace, have organized strong protests against the oil shale industry. In one result, Queensland Energy Resources put the proposed Stuart Oil Shale Project in Australia on hold in 2004.

Petroleum Reservoir

The natural subsurface reservoir is a container of oil, gas, and water where they can move, and its shape is determined by the relationship between the reservoir rock and its surrounding poorly permeable rocks. The subsurface reservoir as only that part of the reservoir rocks where the oil and natural gas can form an accumulation. The natural reservoir remains as such regardless of the type of the contained fluids, or even if it is dry. A comparison can be made with the overpressure. The presence of fluids is necessary for the overpressure. No fluids of different densities, no surplus pressure. As for the reservoirs, they possess two important attributes: spatial limitation (which determines the volume and boundaries of the reservoir), and an internal structure that defines the type and nature of inter-reservoir migration. Indeed, these properties should be included in any definition of the reservoir as well as in the classifications being developed. Some attempts have been made to classify the oil and gas reservoirs on the basis of their relative size (local, zonal, basin-wide, regional, etc.) or their absolute size.

Three types of reservoir limitations can be identified: reservoir roof lateral, and reservoir base. The reservoir roof (top) may cover the reservoir:

1. In normal stratigraphic succession.

2. With some depositional hiatus.

3. May change its age along the strike.

It is improbable but not impossible that a reservoir may be capped by the surface of an impermeable over-thrust. It should be kept in mind the selectivity toward different fluids by caprock and a possibility of its transformation into a reservoir rock during epigenesis. Consequently, it is always important to indicate the exact nature of the fluid. The transformation of the caprock into a reservoir rock results in either disappearance of the reservoir or its conversion into a new reservoir (if there is another caprock above). Lateral limitations are caused by lithologic alterations (including cementation) and the permeability. Small accumulations may be laterally limited by faults. This is possible, but not typical, for the larger reservoirs, because the fault zones in their evolution can become migration paths for various fluids (oil, water, or gas). Fault zones are actually "communication windows" with the other reservoirs. The importance of the presence of the base (bottom) as a necessary reservoir component was not always clearly recognized, because it was believed that the accumulations were formed exclusively by the buoyancy (Archimedes forces). One must always keep in mind that the reservoir is an inseparable part of the hydrodynamic system. This system may be open or with a restricted communication to the surface (artesian), or of "elision" type (with an inverse pattern of hydrostatic pressure). It is not possible for such energy system to exist without a base (bottom).

Reservoir rock is a rock capable of containing oil and gas and yielding them during production. The reservoir rock is characterized by: rock type; permeability type (intergranular, fracture, and combination of the two); the total, intercommunicating, and effective porosity; specific surface area; wettability of rock (oil-wet versus water-wet); fracture type (width, etc.); and fracture distribution. Reservoir is a natural subsurface container for oil, gas, and water. Its existence is predicated on the relationships between the reservoir rock and associated poorly permeable rocks. Reservoir is characterized by reservoir-rock type, relationship with impermeable rocks, reservoir capacity, its hydrodynamic conditions, reservoir energy, and structure. Trap is part of the subsurface reservoir where an oil or gas accumulation can form and be preserved. Its parameters include the reservoir type, reservoir-rock type, conditions of its formation, structure, and capacity. In a special case where the reservoir is lithologically limited from all directions, its parameters may coincide with those of the trap (the entire reservoir is represented by a single trap).

The following features are used in describing a reservoir:

1. Type of the reservoir rock comprising the reservoir.

2. Relationship between the reservoir and the surrounding impermeable rocks.

3. Reservoir capacity.

4. Depositional environment.

In terms of the relationship between the reservoir and its surrounding impermeable rocks, there are three major types of the reservoirs: bedded, massive, and lithologically limited in all directions. Bedded reservoir is a reservoir that is restricted at its top and base by low permeable rocks. The reservoir rock thickness in such a reservoir is more or less or at the edge of the reservoir development, which may result in a pinch-out of the reservoir rock. The reservoir rocks in bedded reservoirs are usually lithologically continuous, but may have a more complex nature. A bedded reservoir may have a single hydrodynamic system. Reservoir energy in bedded reservoirs is distributed in accordance with the hydrostatic or hydrodynamic environment of the artesian basins. However, reservoirs with that kind of energy distribution are typical only for the uppermost portion of the sediment cover. As a result of subsidence and sediment compaction and various secondary geochemical processes, reservoirs may be separated into diverse portions as a consequence of previously described phenomena. Leaving aside changes in the sediment composition, drastic changes occur in the major reservoir-rock properties (porosity and permeability). Even if prior to subsidence the reservoir rocks were reasonably uniform in terms of porosity and permeability, subsequent to subsidence non-uniformities appear between various portions of reservoir so that they may turn out to be totally separated from one another. An indication of such a change may be a change in a hydrodynamic drive from artesian to "elision" type and the appearance of abnormally high pressure.

The beginning of the process involves:

1. Lateral fluid migration, and

2. Gradual change in the reservoir energy.

Potential energy of the accumulations relative to the total energy of the reservoir is small. As the sheet-type reservoir differentiates, lateral migration becomes increasingly more obstructed, with formation of numerous hydraulic fractures. Fluid migration from the reservoir to other favorable zones (if they are available) may become prevalent. An increase in the elastic potential energy is observed (Abnormally-High Formation Pressure, AHFP). Energy distribution becomes discrete.

The difference in potential energy between the accumulations and the reservoir as a whole becomes smaller, and within some zones (blocks) they become identical. Thus, it is reasonable to recognize a separate type, i.e., differentiated sheet-type reservoir, which under certain circumstances becomes a bedded reservoir. Massive reservoir is a thick permeable sequence overlain at the top and restricted from the sides by low-permeable rocks. Its bottom may be at a depth that has not yet been penetrated by wells. Reservoir rocks comprising massive reservoirs may be homogeneous or heterogeneous. Homogeneous massive reservoir rocks may be carbonates and metamorphic or volcanic rocks. Their porosity and permeability is due to the presence of vugs and fractures. Porous and permeable zones in massive reservoir rocks are not stratigraphically related. Isolated high-porosity and high-permeability zones cutting through stratigraphic surfaces within a body of a massif are common. Buried reefs are often assigned to this reservoir type. Among the best examples are the Ishimbay group of fields in Bashkortostan, Russia, and Rainbow Oilfield in Alberta, Canada. Usually, the thickness (height) of massive reservoirs is greater than the width. The length of possible vertical fluid migration is similar or greater than the lateral migration within the beds. The flanks of the reservoir and its contacts with the contemporaneous sediments are steep (thus, the biostromes should be classified as the sheet-type (bedded) rather than the massive-type reservoirs). In as much as the bioherms are very similar to reef buildups, they should be considered as massive reservoirs. To form a trapping mechanism, beside the caprock, the massive reservoirs require isolating steep lateral limitations. As an example, numerous present-day coral reefs in the Indian Ocean do not form subsurface reservoirs not only because of the absence of caprock, but also due to the absence of lateral barriers (lateral isolation). Fresh water accumulating within such bodies floats on the surface of heavier seawater. The hydrodynamic system of the massive reservoirs is poorly studied. It is possible that they communicate at depth with the bedded (sheet-type) reservoirs and are, in effect, just a veriety of a sheet-type reservoir. Reservoirs lithologically limited from all directions include all types of reservoirs where the liquid or gaseous hydrocarbons present from the time of formation of the reservoir are surrounded from all directions with practically impermeable rocks. Fluid movement within such reservoir is limited by its size. There is some superficial similarity between the massive reservoirs and the differentiated sheet-type reservoirs. The similarity is in the limitation (lithologic isolation).

The difference is in the timing of the emergence of the latter. The massive reservoir is a result of depositional processes, whereas the differentiated sheet-type reservoir is a result of stresses during the basin subsidence. The former type is originally small (certainly, any bed is a large lens, but this approach is not used here). The latter type is a separate portion of a previous, possibly large hydrodynamic system. Prevalent elastic energy is typical for both, but the latter type of reservoirs has a greater stress level. The capacity of any type of reservoir is defined by its size and reservoir-rock properties. Energy of a reservoir is associated with its capacity, and the energy is what is important for the extraction of oil and gas (and associated water). The identification of the above four reservoir types is tentative, because such well rooted concepts as reservoir rock and caprock (fluid barrier) are also tentative. Even in the same state, the same rock may be a fluid barrier to one fluid and a reservoir rock to the other, depending on the physiochemical properties of fluids and rocks (especially, wettability), and on the subsurface temperature and pressure. For instance, a prominent projection of a massive reservoir may be just a complication of a regional sheet-type reservoir and be a part of the same hydrodynamic system. This phenomenon is especially common in carbonate sequences. It is possible to imagine a conversion of a sheet-type differentiated (block) reservoir into a massive reservoir if the caprock loses its sealing properties over a fault or flexure. The hydrocarbon accumulation may then occur underneath another, shallower caprock.

Petroleum Refining Processes

Petroleum refining processes are the chemical engineering processes and other facilities used in petroleum refineries (also referred to as oil refineries) to transform crude oil into useful products such as liquefied petroleum gas (LPG), gasoline or petrol, kerosene, jet fuel, diesel oil and fuel oils.

Petroleum refineries are very large industrial complexes that involve many different processing units and auxiliary facilities such as utility units and storage tanks. Each refinery has its own unique arrangement and combination of refining processes largely determined by the refinery location, desired products and economic considerations.

Some modern petroleum refineries process as much as 800,000 to 900,000 barrels (127,000 to 143,000 cubic meters) per day of crude oil.

Petroleum refinery in Anacortes, Washington, United States.

Prior to the nineteenth century, petroleum was known and utilized in various fashions in Babylon, Egypt, China, Philippines, Rome and along the Caspian Sea. However, the modern history of the petroleum industry is said to have begun in 1846 when Abraham Gessner of Nova Scotia, Canada devised a process to produce kerosene from coal. Shortly thereafter, in 1854, Ignacy Lukasiewicz began producing kerosene from hand-dug oil wells near the town of Krosno, Poland. The first large petroleum refinery was built in Ploesti, Romania in 1856 using the abundant oil available in Romania.

In North America, the first oil well was drilled in 1858 by James Miller Williams in Ontario, Canada. In the United States, the petroleum industry began in 1859 when Edwin Drake found oil near Titusville, Pennsylvania. The industry grew slowly in the 1800s, primarily producing kerosene for oil lamps. In the early twentieth century, the introduction of the internal combustion engine and its use in automobiles created a market for gasoline that was the impetus for fairly rapid growth of the petroleum industry. The early finds of petroleum like those in Ontario and Pennsylvania were soon outstripped by large oil "booms" in Oklahoma, Texas and California.

Prior to World War II in the early 1940s, most petroleum refineries in the United States consisted simply of crude oil distillation units (often referred to as atmospheric crude oil distillation units). Some refineries also had vacuum distillation units as well as thermal cracking units such as visbreakers (viscosity breakers, units to lower the viscosity of the oil). All of the many other refining processes discussed below were developed during the war or within a few years after the war. They became commercially available within 5 to 10 years after the war ended and the worldwide petroleum industry experienced very rapid growth. The driving force for that growth in technology and in the number and size of refineries worldwide was the growing demand for automotive gasoline and aircraft fuel.

In the United States, for various complex economic and political reasons, the construction of new refineries came to a virtual stop in about the 1980s. However, many of the existing refineries in the United States have revamped many of their units and constructed add-on units in order to: increase their crude oil processing capacity, increase the octane rating of their product gasoline, lower the sulfur content of their diesel fuel and home heating fuels to comply with environmental regulations and comply with environmental air pollution and water pollution requirements.

Processing Units used in Refineries

- Crude Oil Distillation unit: Distills the incoming crude oil into various fractions for further processing in other units.

- Vacuum distillation unit: Further distills the residue oil from the bottom of the crude oil distillation unit. The vacuum distillation is performed at a pressure well below atmospheric pressure.

- Naphtha hydrotreater unit: Uses hydrogen to desulfurize the naphtha fraction from the crude oil distillation or other units within the refinery.

- Catalytic reforming unit: Converts the desulfurized naphtha molecules into higher-octane molecules to produce *reformate*, which is a component of the end-product gasoline or petrol.

- Alkylation unit: Converts isobutane and butylenes into *alkylate*, which is a very high-octane component of the end-product gasoline or petrol.

- Isomerization unit: Converts linear molecules such as normal pentane into higher-octane branched molecules for blending into the end-product gasoline. Also used to convert linear normal butane into isobutane for use in the alkylation unit.

- Distillate hydrotreater unit: Uses hydrogen to desulfurize some of the other distilled fractions from the crude oil distillation unit (such as diesel oil).

- Merox (mercaptan oxidizer) or similar units: Desulfurize LPG, kerosene or jet fuel by oxidizing undesired mercaptans to organic disulfides.

- Amine gas treater, Claus unit, and tail gas treatment for converting hydrogen sulfide gas from the hydrotreaters into end-product elemental sulfur. The large majority of the 64,000,000 metric tons of sulfur produced worldwide in 2005 was byproduct sulfur from petroleum refining and natural gas processing plants.

- Fluid catalytic cracking (FCC) unit: Upgrades the heavier, higher-boiling fractions from the crude oil distillation by converting them into lighter and lower boiling, more valuable products.

- Hydrocracker unit: Uses hydrogen to upgrade heavier fractions from the crude oil distillation and the vacuum distillation units into lighter, more valuable products.

- Visbreaker unit upgrades heavy residual oils from the vacuum distillation unit by thermally cracking them into lighter, more valuable reduced viscosity products.

- Delayed coking and fluid coker units: Convert very heavy residual oils into end-product petroleum coke as well as naphtha and diesel oil by-products.

Auxiliary Facilities Required in Refineries

- Steam reforming unit: Converts natural gas into hydrogen for the hydrotreaters and the hydrocracker.

- Sour water stripper unit: Uses steam to remove hydrogen sulfide gas from various wastewater streams for subsequent conversion into end-product sulfur in the Claus unit.

- Utility units such as cooling towers for furnishing circulating cooling water, steam generators, instrument air systems for pneumatically operated control valves and an electrical substation.

- Wastewater collection and treating systems consisting of API separators, dissolved air flotation (DAF) units and some type of further treatment (such as an activated sludge biotreater) to make the wastewaters suitable for reuse or for disposal.

- Liquified gas (LPG) storage vessels for propane and similar gaseous fuels at a pressure sufficient to maintain them in liquid form. These are usually spherical vessels or *bullets* (horizontal vessels with rounded ends).

- Storage tanks for crude oil and finished products, usually vertical, cylindrical vessels with some sort of vapour emission control and surrounded by an earthen berm to contain liquid spills.

The Crude Oil Distillation Unit

The crude oil distillation unit (CDU) is the first processing unit in virtually all petroleum refineries. The CDU distills the incoming crude oil into various fractions of different boiling ranges, each of which are then processed further in the other refinery processing units. The CDU is often referred to as the *atmospheric distillation unit* because it operates at slightly above atmospheric pressure.

Below is a schematic flow diagram of a typical crude oil distillation unit. The incoming crude oil is preheated by exchanging heat with some of the hot, distilled fractions and other streams. It is then desalted to remove inorganic salts (primarily sodium chloride).

Following the desalter, the crude oil is further heated by exchanging heat with some of the hot, distilled fractions and other streams. It is then heated in a fuel-fired furnace (fired heater) to a temperature of about 398 °C and routed into the bottom of the distillation unit.

The cooling and condensing of the distillation tower overhead is provided partially by exchanging heat with the incoming crude oil and partially by either an air-cooled or water-cooled condenser. Additional heat is removed from the distillation column by a pumparound system as shown in the diagram below.

As shown in the flow diagram, the overhead distillate fraction from the distillation column is naphtha. The fractions removed from the side of the distillation column at various points between the column top and bottom are called *sidecuts*. Each of the sidecuts (i.e., the kerosene, light gas oil and heavy gas oil) is cooled by exchanging heat with the incoming crude oil. All of the fractions (i.e., the overhead naphtha, the sidecuts and the bottom residue) are sent to intermediate storage tanks before being processed further.

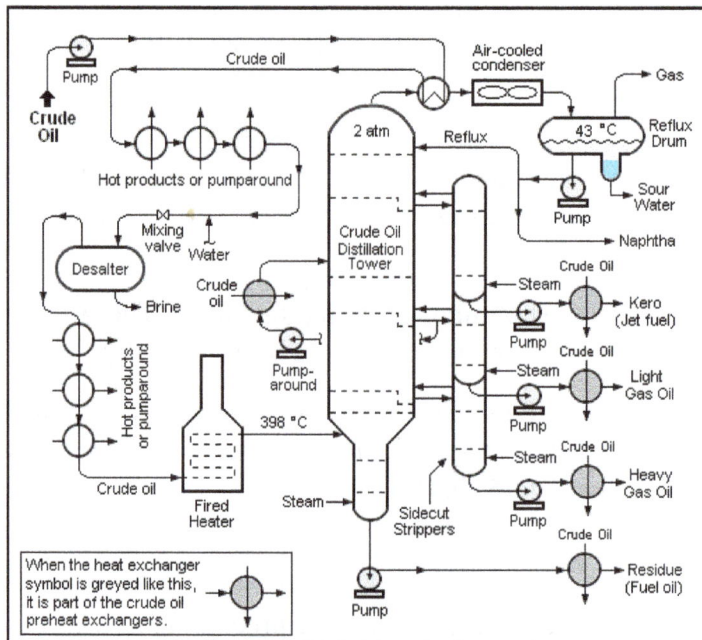

Schematic flow diagram of a typical crude oil distillation unit
as used in petroleum crude oil refineries.

Flow Diagram of a Typical Petroleum Refinery

The image below is a schematic flow diagram of a typical petroleum refinery that depicts the various refining processes and the flow of intermediate product streams that occurs between the inlet crude oil feedstock and the final end-products.

The diagram depicts only one of the literally hundreds of different oil refinery configurations. The diagram also does not include any of the usual refinery facilities providing utilities such as steam, cooling water, and electric power as well as storage tanks for crude oil feedstock and for intermediate products and end products.

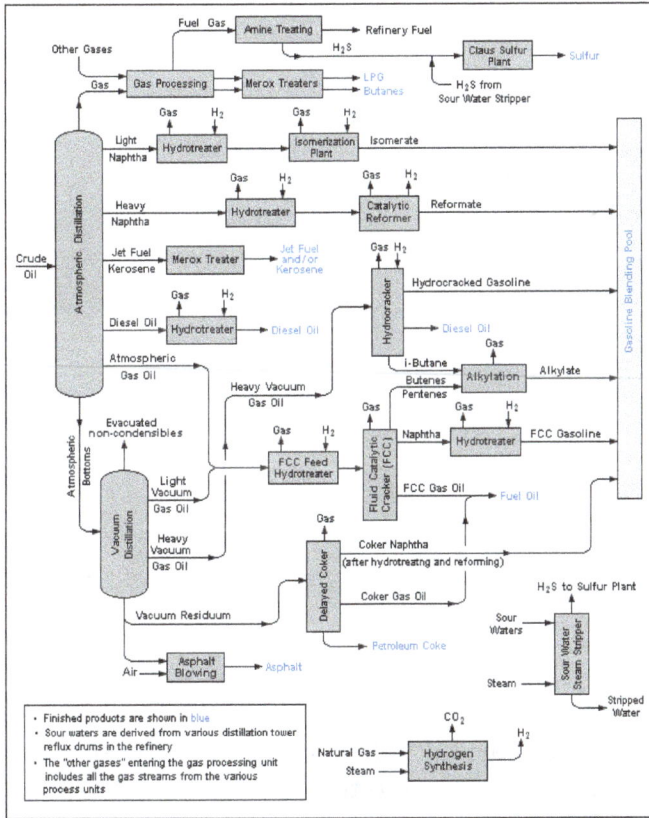

A schematic flow diagram of a typical petroleum refinery.

Refining End-products

The primary end-products produced in petroleum refining may be grouped into four categories: light distillates, middle distillates, heavy distillates and others.

Light Distillates

- C1 and C2 components.

- Liquified petroleum gas (LPG).

- Light naphtha.

- Gasoline (petrol).

- Heavy naphtha.

Middle Distillates

- Kerosene.

- Automotive and rail-road diesel fuels.

- Residential heating fuel.

- Other light fuel oils.

Heavy Distillates

- Heavy fuel oils.

- Wax.

- Lubricating oils.

- Asphalt.

Others

- Coke (similar to coal).

Oil Well

An oil well is a hole dug into the Earth that serves the purpose of bringing oil or other hydrocarbons - such as natural gas - to the surface. Oil wells almost always produce some natural gas and frequently bring water up with the other petroleum products.

Types of Wells

There are numerous different ways that oil well can be drilled to maximize the output of the well while minimizing other costs. The most common type of well drilled today is known as a conventional well. These wells are wells drilled in the traditional sense in that a location is chosen above the reservoir and the well is drilled vertically downward. Additionally, wells with a small amount of deviation in their path from the vertical are also considered to be conventional. This slight turning of the well is obtained during drilling by using a type of steerable device that shifts the direction the well is being dug. These wells are the most common and are fairly inexpensive to drill.

Horizontal wells are an alternative type of well used when conventional wells do not yield enough fuel. These wells are drilled and steered to enter a deposit nearly horizontally. These wells can hit targets and stimulate reservoirs in ways that a vertical well cannot. Combined with hydraulic fracturing previously unproductive rocks can be used as sources for natural gas. Examples of these types of deposits include formations that contain shale gas or tight gas.

Other types of wells include offshore wells, which are wells that are drilled in the water instead of onshore. These provide access to previous inaccessible oil deposits. Multilateral wells are wells used occasionally that have several branches off of the main borehole that drain a separate part of the reservoir.

Drilling a Well

To drill a well, a specialized piece of equipment known as a drilling rig bores a hole through many layers of dirt and rock until it reaches the oil and gas reservoir where the oil is held. The size of the borehole differs from well to well, but is generally around 12.5 to 90 centimeters wide. To cut through this rock, the drill is pushed down by the weight of the piping above it. This piping is used to pump a thick fluid known as mud into the well. This mud assists in the drilling process by maintaining the pressure below ground in the well as well as by collecting debris created from the drilling and bringing it up to the surface. As the drill digs deeper sections of piping are attached to lengthen the well.

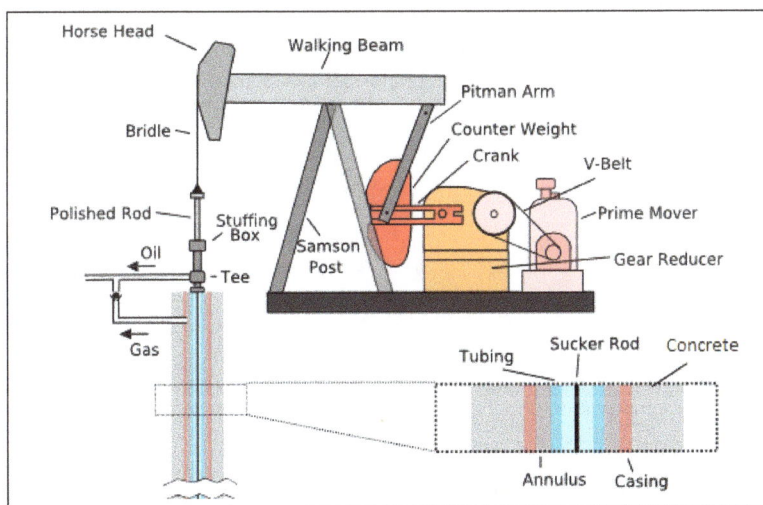

Diagram of an oil well showing how it connects to a pump jack.

After the well is drilled it is completed and cased. The well casing is the lining that is inserted between the edge of the well and the well itself that helps to structurally support the well. In a closed hole well, concrete is poured into the space between the pipe and the borehole for stability and to prevent groundwater contamination from seepage. Once casing is complete, the top of the well is fixed with a production tree - a set of valves that controls the flow of oil out of the well and the pressure within the well.

References

- Alan S Morris (9 March 2001). Measurement and Instrumentation Principles. Butterworth-Heinemann. Pp. 328–. ISBN 978-0-08-049648-1

- Oil-industry, definition: petropedia.com, Retrieved 29 March, 2019

- Madani, Ismail M.; Khalfan, Sameer; Khalfan, Hussain; Jidah, Jasim; Nabeel Aladin, M. (1992-04-01). "Occupational exposure to carbon monoxide during charcoal meat grilling". Science of the Total Environment. 114: 141–147. Bibcode:1992scten.114..141M. Doi:10.1016/0048-9697(92)90420-W. ISSN 0048-9697

- Oil-gas-industry-overview, investing: investopedia.com, Retrieved 30 April, 2019

- "33 accidents happened at oil refineries as EPA delayed updating disaster rule, says environmentalist group". Daily Breeze. 2018-04-04. Retrieved 2018-12-11

- What-is-petroleum-reservoir-and-how; geologylearn.blogspot.com, Retrieved 1 May, 2019

- Salim Al-Hassani (2008). "1000 Years of Missing Industrial History". In Emilia Calvo Labarta; Mercè Comes Maymo; Roser Puig Aguilar; Mònica Rius Pinies (eds.). A shared legacy: Islamic science East and West. Edicions Universitat Barcelona. Pp. 57–82 [63]. ISBN 84-475-3285-2

- Oil-well, encyclopedia: energyeducation.ca, Retrieved 2 June, 2019

5

Byproducts of Petroleum Production

Some of the most common byproducts of petroleum engineering are paraffin wax, petrochemicals, liquefied petroleum gas, petroleum coke, crude oil, diesel fuel, petroleum ether, petroleum jelly, gasoline and kerosene. This chapter has been carefully written to provide an easy understanding of these byproducts of petroleum production.

Paraffin Wax

Paraffin wax is a colourless or white, somewhat translucent, hard wax consisting of a mixture of solid straight-chain hydrocarbons ranging in melting point from about 48° to 66° C (120° to 150 °F). Paraffin wax is obtained from petroleum by dewaxing light lubricating oil stocks. It is used in candles, wax paper, polishes, cosmetics, and electrical insulators. It assists in extracting perfumes from flowers, forms a base for medical ointments, and supplies a waterproof coating for wood. In wood and paper matches, it helps to ignite the matchstick by supplying an easily vaporized hydrocarbon fuel.

Paraffin wax was first produced commercially in 1867, less than 10 years after the first petroleum well was drilled. Paraffin wax precipitates readily from petroleum on chilling. Technical progress has served only to make the separations and filtration more efficient and economical. Purification methods consist of chemical treatment, decolorization by adsorbents, and fractionation of the separated waxes into grades by distillation, recrystallization, or both. Crude oils differ widely in wax content.

Synthetic paraffin wax was introduced commercially after World War II as one of the products obtained in the Fischer–Tropsch reaction, which converts coal gas to hydrocarbons. Snow-white and harder than petroleum paraffin wax, the synthetic product has a unique character and high purity that make it a suitable replacement for certain vegetable waxes and as a modifier for petroleum waxes and for some plastics, such as polyethylene. Synthetic paraffin waxes may be oxidized to yield pale-yellow, hard waxes of high molecular weight that can be saponified with aqueous solutions of organic or inorganic alkalies, such as borax, sodium hydroxide, triethanolamine, and morpholine. These wax dispersions serve as heavy-duty floor wax, as waterproofing for textiles and paper, as tanning agents for leather, as metal-drawing lubricants, as rust preventives, and for masonry and concrete treatment.

Petrochemicals

Petrochemicals (also known as petroleum distillates) are chemical products derived from petroleum. Some chemical compounds made from petroleum are also obtained from other fossil fuels, such as coal or natural gas, or renewable sources such as corn, palm fruit or sugar cane.

The two most common petrochemical classes are olefins (including ethylene and propylene) and aromatics (including benzene, toluene and xylene isomers).

Oil refineries produce olefins and aromatics by fluid catalytic cracking of petroleum fractions. Chemical plants produce olefins by steam cracking of natural gas liquids like ethane and propane. Aromatics are produced by catalytic reforming of naphtha. Olefins and aromatics are the building-blocks for a wide range of materials such as solvents, detergents, and adhesives. Olefins are the basis for polymers and oligomers used in plastics, resins, fibers, elastomers, lubricants, and gels.

Global ethylene and propylene production are about 115 million tonnes and 70 million tonnes per annum, respectively. Aromatics production is approximately 70 million tonnes. The largest petrochemical industries are located in the USA and Western Europe; however, major growth in new production capacity is in the Middle East and Asia. There is substantial inter-regional petrochemical trade.

Primary petrochemicals are divided into three groups depending on their chemical structure:

- Olefins includes Ethene, Propene, Butenes and butadiene. Ethylene and propylene are important sources of industrial chemicals and plastics products. Butadiene is used in making synthetic rubber.

- Aromatics includes Benzene, toluene and xylenes, as a whole referred to as BTX and primarily obtained from petroleum refineries by extraction from the reformate produced in catalytic reformers using Naphtha obtained from petroleum refineries. Benzene is a raw material for dyes and synthetic detergents, and benzene and toluene for isocyanates MDI and TDI used in making polyurethanes. Manufacturers use xylenes to produce plastics and synthetic fibers.

- Synthesis gas is a mixture of carbon monoxide and hydrogen used to make ammonia and methanol. Ammonia is used to make the fertilizer urea and methanol is used as a solvent and chemical intermediate. Steam crackers are not to be confused with steam reforming plants used to produce hydrogen and ammonia.

- Methane, ethane, propane and butanes obtained primarily from natural gas processing plants.

- Methanol and formaldehyde.

In 2007, the amounts of ethylene and propylene produced in steam crackers were about 115 Mt (megatonnes) and 70 Mt, respectively. The output ethylene capacity of large steam crackers ranged up to as much as 1.0 – 1.5 Mt per year.

The adjacent diagram schematically depicts the major hydrocarbon sources and processes used in producing petrochemicals.

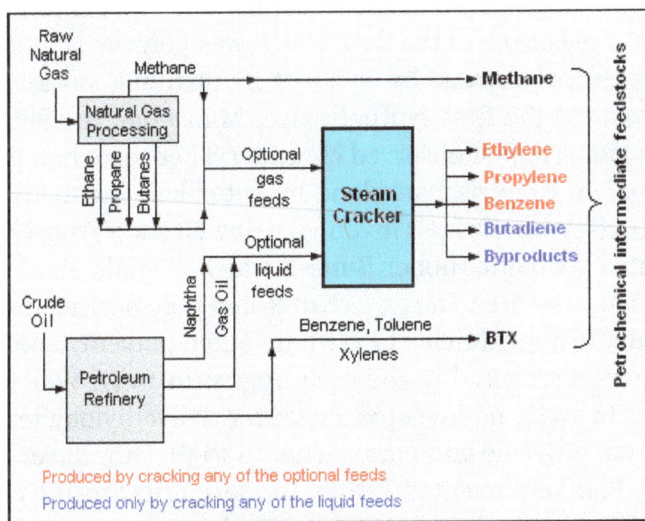

Petrochemical feedstock sources.

Like commodity chemicals, petrochemicals are made on a very large scale. Petrochemical manufacturing units differ from commodity chemical plants in that they often produce a number of related products. Compare this with specialty chemical and fine chemical manufacture where products are made in discrete batch processes.

Petrochemicals are predominantly made in a few manufacturing locations around the world, for example in Jubail & Yanbu Industrial Cities in Saudi Arabia, Texas & Louisiana in the US, in Teesside in the Northeast of England in the United Kingdom, in Rotterdam in the Netherlands, and in Jamnagar & Dahej in Gujarat, India. Not all of the petrochemical or commodity chemical materials produced by the chemical industry are made in one single location but groups of related materials are often made in adjacent manufacturing plants to induce industrial symbiosis as well as material and utility efficiency and other economies of scale. This is known in chemical engineering terminology as integrated manufacturing. Speciality and fine chemical companies are sometimes found in similar manufacturing locations as petrochemicals but, in most cases, they do not need the same level of large scale infrastructure (e.g., pipelines, storage, ports and power, etc.) and therefore can be found in multi-sector business parks.

The large scale petrochemical manufacturing locations have clusters of manufacturing units that share utilities and large scale infrastructure such as power stations, storage tanks, port facilities, road and rail terminals. In the United Kingdom for example, there are 4 main locations for such manufacturing: near the River Mersey in Northwest England, on the Humber on the East coast of Yorkshire, in Grangemouth near the Firth of Forth in Scotland and in Teesside as part of the Northeast of England Process Industry Cluster (NEPIC). To demonstrate the clustering and integration, some 50% of the United Kingdom's petrochemical and commodity chemicals are produced by the NEPIC industry cluster companies in Teesside.

In 1835, Henri Victor Regnault, a French chemist left vinyl chloride in the sun and found white solid at the bottom of the flask which was polyvinyl chloride. In 1839 Eduard Simon, discovered polystyrene by accident by distilling storax. In 1856, William Henry Perkin discovered the first synthetic dye, Mauveine. In 1888, Friedrich Reinitzer, an Austrian plant scientist observed cholesteryl benzoate had two different melting points. In 1909, Leo Hendrik Baekeland invented bakelite made from phenol and formaldehyde. In 1928 synthetic fuels invented using Fischer-Tropsch process. In 1929, Walter Bock invented synthetic rubber Buna-S which is made up of styrene and butadiene and used to make car tires. In 1933, Otto Röhm polymerized the first acrylic glass methyl methacrylate. In 1935, Michael Perrin invented polyethylene. After World War II, polypropylene was discovered in the early 1950s. In 1937, Wallace Hume Carothers invented nylon. In 1946, he invented Polyester. Polyethylene terephthalate (PET) bottles are made from ethylene and paraxylene. In 1938, Otto Bayer invented polyurethane. In 1941, Roy Plunkett invented Teflon. In 1949, Fritz Stastny turned polystyrene into foam. In 1965, Stephanie Kwolek invented Kevlar.

Olefins

The following is a partial list of the major commercial petrochemicals and their derivatives:

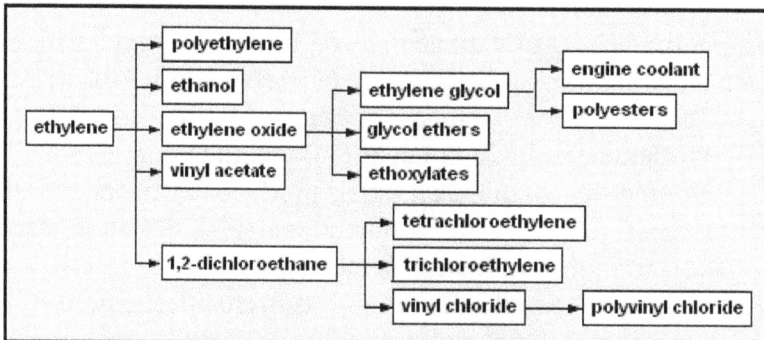

Chemicals produced from ethylene:

- Ethylene: The simplest olefin; used as a chemical feedstock and ripening stimulant:

  ○ Polyethylene: Polymerized ethylene; LDPE, HDPE, LLDPE.

  ○ Ethanol: Via ethylene hydration (chemical reaction adding water) of ethylene.

  ○ Ethylene oxide: Via ethylene oxidation.

    ▪ Ethylene glycol: Via ethylene oxide hydration.

      ▫ Engine coolant: Ethylene glycol, water and inhibitor mixture.

      ▫ Polyesters: Any of several polymers with ester linkages in the main chain.

    ▪ Glycol ethers: via glycol condescension.

    ▪ Ethoxylates.

  ○ Vinyl Acetate.

  ○ 1, 2-dichloroethane:

    ▪ Trichloroethylene.

    ▪ Tetrachloroethylene: Also called perchloroethylene; used as a dry cleaning solvent and degreaser.

    ▪ Vinyl chloride: Monomer for polyvinyl chloride.

      ▫ Polyvinyl chloride (PVC): Type of plastic used for piping, tubing, other things.

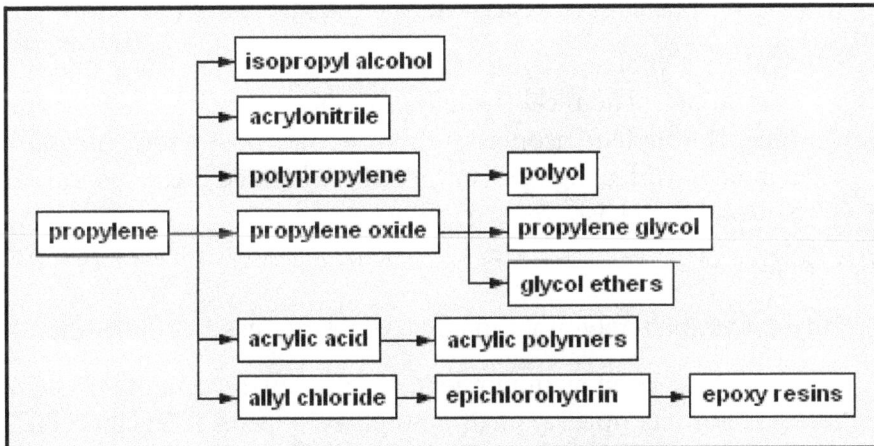

Chemicals produced from propylene.

- Propylene: Used as a monomer and a chemical feedstock.

  ○ Isopropyl alcohol: 2-propanol; often used as a solvent or rubbing alcohol.

  ○ Acrylonitrile: Useful as a monomer in forming Orlon, ABS.

  ○ Polypropylene: Polymerized propylene.

  ○ Propylene oxide:

    ▪ Polyether polyol: Used in the production of polyurethanes.

    ▪ Propylene glycol: Used in engine coolant and aircraft deicer fluid.

    ▪ Glycol ethers: From condensation of glycols.

  ○ Acrylic Acid:

    ▪ Acrylic polymers.

  ○ Allyl chloride:

    ▪ Epichlorohydrin: Chloro-oxirane; used in epoxy resin formation.

      ▫ Epoxy resins: A type of polymerizing glue from bisphenol A, epichlorohydrin, and some amine.

- Butene:

  ○ Isomers of butylene: Useful as monomers or co-monomers.

    ▪ Isobutylene: Feed for making methyl *tert*-butyl ether (MTBE) or monomer for copolymerization with a low percentage of isoprene to make butyl rubber.

  ○ 1,3-butadiene (or buta-1,3-diene): A diene often used as a monomer or co-monomer for polymerization to elastomers such as polybutadiene, styrene-butadiene rubber, or a plastic such as acrylonitrile-butadiene-styrene (ABS).

    ▪ Synthetic rubbers: Synthetic elastomers made of any one or more of several petrochemical (usually) monomers such as 1,3-butadiene, styrene, isobutylene, isoprene, chloroprene; elastomeric polymers are often made with a high percentage of conjugated diene monomers such as 1,3-butadiene, isoprene, or chloroprene.

- Higher olefins:

  ○ Polyolefins such poly-alpha-olefins, which are used as lubricants.

  ○ Alpha-olefins: Used as monomers, co-monomers, and other chemical precursors. For example, a small amount of 1-hexene can be copolymerized with ethylene into a more flexible form of polyethylene.

- ◦ Other higher olefins.

- ◦ Detergent alcohols.

Aromatics

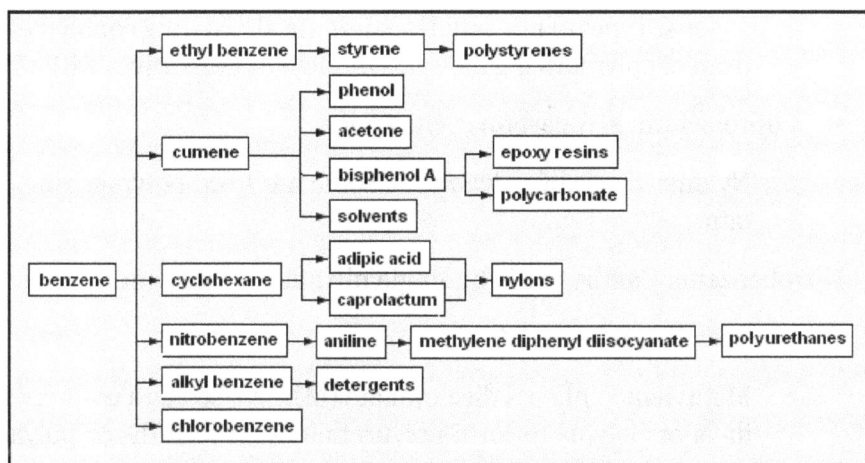

Chemicals produced from benzene.

- • Benzene: The simplest aromatic hydrocarbon.

  - ◦ Ethylbenzene: Made from benzene and ethylene.

    - ▪ Styrene made by dehydrogenation of ethylbenzene; used as a monomer:

      - ▫ Polystyrenes: Polymers with styrene as a monomer.

  - ◦ Cumene: Isopropylbenzene; a feedstock in the cumene process.

    - ▪ Phenol: Hydroxybenzene; often made by the cumene process.

    - ▪ Acetone: Dimethyl ketone; also often made by the cumene process.

    - ▪ Bisphenol A: A type of "double" phenol used in polymerization in epoxy resins and making a common type of polycarbonate.

      - ▫ Epoxy resins: A type of polymerizing glue from bisphenol A, epichlorohydrin, and some amine.

      - ▫ Polycarbonate: A plastic polymer made from bisphenol A and phosgene (carbonyl dichloride).

    - ▪ Solvents: Liquids used for dissolving materials; examples often made from petrochemicals include ethanol, isopropyl alcohol, acetone, benzene, toluene, xylenes.

  - ◦ Cyclohexane: A 6-carbon aliphatic cyclic hydrocarbon sometimes used as a non-polar solvent.

- Adipic acid: A 6-carbon dicarboxylic acid, which can be a precursor used as a co-monomer together with a diamine to form an alternating copolymer form of nylon.

  □ Nylons: Types of polyamides, some are alternating copolymers formed from copolymerizing dicarboxylic acid or derivatives with diamines.

- Caprolactam: A 6-carbon cyclic amide.

  □ Nylons: Types of polyamides, some are from polymerizing caprolactam.

○ Nitrobenzene: Can be made by single nitration of benzene.

- Aniline: Aminobenzene.

  □ Methylene diphenyl diisocyanate (MDI): Used as a co-monomer with diols or polyols to form polyurethanes or with di- or polyamines to form polyureas.

○ Alkylbenzene: A general type of aromatic hydrocarbon, which can be used as a precursor for a sulfonate surfactant (detergent).

- Detergents: Often include surfactants types such as alkylbenzenesulfonates and nonylphenol ethoxylates.

○ Chlorobenzene.

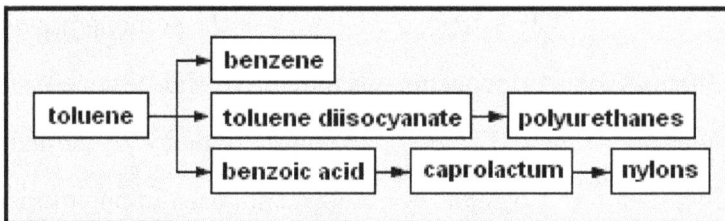

Chemicals produced from toluene.

- Toluene: Methylbenzene; can be a solvent or precursor for other chemicals.

○ Benzene.

○ Toluene diisocyanate (TDI): Used as co-monomers with polyether polyols to form polyurethanes or with di- or polyamines to form polyureas polyurethanes.

○ Benzoic acid: Carboxybenzene.

- Caprolactam.

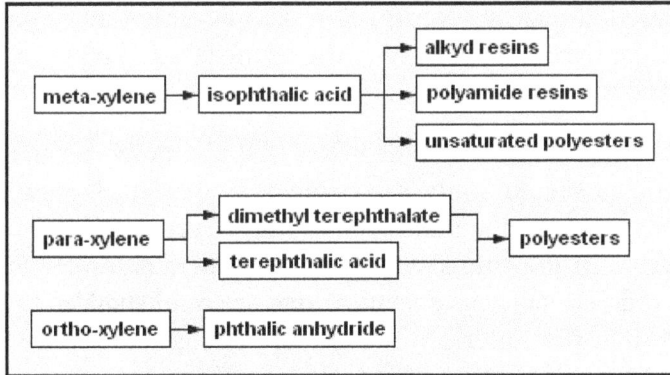

Chemicals produced from xylenes.

- Mixed xylenes: Any of three dimethylbenzene isomers, could be a solvent but more often precursor chemicals.

  - *Ortho*-xylene: Both methyl groups can be oxidized to form (*ortho*-)phthalic acid.

    - Phthalic anhydride.

  - *Para*-xylene: Both methyl groups can be oxidized to form terephthalic acid.

    - Dimethyl terephthalate: Can be copolymerized to form certain polyesters.

      - Polyesters: Although there can be many types, polyethylene terephthalate is made from petrochemical products and is very widely used.

    - Purified terephthalic acid: Often copolymerized to form polyethylene terephthalate.

      - Polyesters.

  - *Meta*-xylene:

    - Isophthalic acid:

      - Alkyd resins.

      - Polyamide resins.

      - Unsaturated polyesters.

Liquefied Petroleum Gas

Liquefied petroleum gas or liquid petroleum gas (LPG or LP gas), also referred to as simply propane or butane, are flammable mixtures of hydrocarbon gases used as fuel in heating appliances, cooking equipment, and vehicles.

It is increasingly used as an aerosol propellant and a refrigerant, replacing chlorofluorocarbons in an effort to reduce damage to the ozone layer. When specifically used as a vehicle fuel it is often referred to as autogas.

Varieties of LPG bought and sold include mixes that are mostly propane (C_3H_8), mostly butane (C_4H_{10}), and, most commonly, mixes including both propane and butane. In the northern hemisphere winter, the mixes contain more propane, while in summer, they contain more butane. In the United States, mainly two grades of LPG are sold: commercial propane and HD-5. These specifications are published by the Gas Processors Association (GPA) and the American Society of Testing and Materials (ASTM). Propane/butane blends are also listed in these specifications.

Propylene, butylenes and various other hydrocarbons are usually also present in small concentrations. HD-5 limits the amount of propylene that can be placed in LPG to 5%, and is utilized as an autogas specification. A powerful odorant, ethanethiol, is added so that leaks can be detected easily. The internationally recognized European Standard is EN 589. In the United States, tetrahydrothiophene (thiophane) or amyl mercaptan are also approved odorants, although neither is currently being utilized.

LPG is prepared by refining petroleum or "wet" natural gas, and is almost entirely derived from fossil fuel sources, being manufactured during the refining of petroleum (crude oil), or extracted from petroleum or natural gas streams as they emerge from the ground. It was first produced in 1910 by Dr. Walter Snelling, and the first commercial products appeared in 1912. It currently provides about 3% of all energy consumed, and burns relatively cleanly with no soot and very few sulfur emissions. As it is a gas, it does not pose ground or water pollution hazards, but it can cause air pollution. LPG has a typical specific calorific value of 46.1 MJ/kg compared with 42.5 MJ/kg for fuel oil and 43.5 MJ/kg for premium grade petrol (gasoline). However, its energy density per volume unit of 26 MJ/L is lower than either that of petrol or fuel oil, as its relative density is lower (about 0.5–0.58 kg/L, compared to 0.71–0.77 kg/L for gasoline).

As its boiling point is below room temperature, LPG will evaporate quickly at normal temperatures and pressures and is usually supplied in pressurised steel vessels. They are typically filled to 80–85% of their capacity to allow for thermal expansion of the contained liquid. The ratio between the volumes of the vaporized gas and the liquefied gas varies depending on composition, pressure, and temperature, but is typically around 250:1. The pressure at which LPG becomes liquid, called its vapour pressure, likewise varies depending on composition and temperature; for example, it is approximately 220 kilopascals (32 psi) for pure butane at 20 °C (68 °F), and approximately 2,200 kilopascals (320 psi) for pure propane at 55 °C (131 °F). LPG is heavier than air, unlike natural gas, and thus will flow along floors and tend to settle in low spots, such as basements. There are two main dangers from this. The first is a possible explosion if the mixture of LPG and air is within the explosive limits and there is an ignition source. The second is suffocation due to LPG displacing air, causing a decrease in oxygen concentration.

A "full" LPG cylinder contains 85% liquid, the ullage volume will contain vapour at a pressure that varies with temperature.

Two 45 kg (99 lb) LPG cylinders in New Zealand used for domestic supply.

A group of 25 kg (55 lb) LPG cylinders in Malta.

LPG minibuses in Hong Kong.

LPG Ford Falcon taxicab in Perth.

Tank cars in a Canadian train for carrying liquefied petroleum gas by rail.

Uses

LPG has a very wide variety of uses, mainly used for cylinders across many different markets as an efficient fuel container in the agricultural, recreation, hospitality, industrial, construction, sailing and fishing sectors. It can serve as fuel for cooking, central heating and to water heating and is a particularly cost-effective and efficient way to heat off-grid homes.

Cooking

LPG is used for cooking in many countries for economic reasons, for convenience or because it is the preferred fuel source.

In India, nearly 8.9 million tons of LPG was consumed in the six months between April and September 2016 in the domestic sector, mainly for cooking. The number of domestic connections are 215 million (i.e., one connection for every six people) with a circulation of more than 350 million LPG cylinders. Most of the LPG requirement is imported. Piped city gas supply in India is not yet developed on major scale. LPG is subsidised by the Indian government for domestic users. Increase in LPG prices has been a politically sensitive matter in India as it potentially affects the middle class voting pattern.

LPG was once a standard cooking fuel in Hong Kong; however, the continued expansion of town gas to newer buildings has reduced LPG usage to less than 24% of residential units. However, other than electric, induction, or infrared stoves, LPG-fueled stoves are the only type available in most suburban villages and many public housing estates.

LPG is the most common cooking fuel in Brazilian urban areas, being used in virtually all households, with the exception of the cities of Rio de Janeiro and São Paulo, which have a natural gas pipeline infrastructure. Since 2001, poor families receive a government grant ("Vale Gás") used exclusively for the acquisition of LPG. Since 2003, this grant is part of the government main social welfare program ("Bolsa Família"). Also, since 2005 the national oil company Petrobras differentiates between LPG destined for cooking and LPG destined for other uses, practicing a lower price for the former. This is a result of a directive from Brazilian federal government, but its demise is currently being debated.

LPG is commonly used in North America for domestic cooking and outdoor grilling.

Rural Heating

LPG Cylinders in India.

Predominantly in Europe and rural parts of many countries, LPG can provide an alternative to electric heating, heating oil, or kerosene. LPG is most often used in areas that do not have direct access to piped natural gas.

LPG can be used as a power source for combined heat and power technologies (CHP). CHP is the process of generating both electrical power and useful heat from a single fuel source. This technology has allowed LPG to be used not just as fuel for heating and cooking, but also for decentralized generation of electricity.

LPG can be stored in a variety of manners. LPG, as with other fossil fuels, can be combined with renewable power sources to provide greater reliability while still achieving some reduction in CO_2 emissions. However, as opposed to wind and solar renewable energy sources, LPG can be used as a standalone energy source without the prohibitive expense of electrical energy storage. In many climates renewable sources such as solar and wind power would still require the construction, installation and maintenance of reliable baseload power sources such as LPG fueled generation to provide electrical power during the entire year. 100% wind/solar is possible, the caveat being that even in 2018 the expense of the additional generation capacity necessary to charge batteries, plus the cost of battery electrical storage, still makes this option economically feasible in only a minority of situations.

Motor Fuel

When LPG is used to fuel internal combustion engines, it is often referred to as autogas or auto propane. In some countries, it has been used since the 1940s as a petrol alternative for spark ignition engines. In some countries, there are additives in the liquid that extend engine life and the ratio of butane to propane is kept quite precise in fuel LPG. Two recent studies have examined LPG-fuel-oil fuel mixes and found that smoke emissions and fuel consumption are reduced but hydrocarbon emissions are increased. The studies were split on CO emissions, with one finding significant increases, and the other finding slight increases at low engine load but a considerable decrease at high engine load. Its advantage is that it is non-toxic, non-corrosive and free of tetraethyl-lead or any additives, and has a high octane rating (102–108 RON depending on local specifications). It burns more cleanly than petrol or fuel-oil and is especially free of the particulates present in the latter.

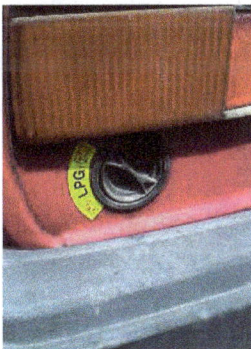

LPG filling connector on a car.

White bordered green diamond symbol used on LPG-powered vehicles in China.

LPG has a lower energy density per liter than either petrol or fuel-oil, so the equivalent fuel consumption is higher. Many governments impose less tax on LPG than on petrol or fuel-oil, which helps offset the greater consumption of LPG than of petrol or fuel-oil. However, in many European countries this tax break is often compensated by a much higher annual tax on cars using LPG than on cars using petrol or fuel-oil. Propane is the third most widely used motor fuel in the world. 2013 estimates are that over 24.9

million vehicles are fueled by propane gas worldwide. Over 25 million tonnes (over 9 billion US gallons) are used annually as a vehicle fuel.

Not all automobile engines are suitable for use with LPG as a fuel. LPG provides less upper cylinder lubrication than petrol or diesel, so LPG-fueled engines are more prone to valve wear if they are not suitably modified. Many modern common rail diesel engines respond well to LPG use as a supplementary fuel. This is where LPG is used as fuel as well as diesel. Systems are now available that integrate with OEM engine management systems.

Conversion to Gasoline

LPG can be converted into alkylate which is a premium gasoline blending stock because it has exceptional anti-knock properties and gives clean burning.

Refrigeration

LPG is instrumental in providing off-the-grid refrigeration, usually by means of a gas absorption refrigerator.

Blended of pure, dry propane (refrigerant designator R-290) and isobutane (R-600a) the blend "R-290a" has negligible ozone depletion potential and very low global warming potential and can serve as a functional replacement for R-12, R-22, R-134a and other chlorofluorocarbon or hydrofluorocarbon refrigerants in conventional stationary refrigeration and air conditioning systems.

Such substitution is widely prohibited or discouraged in motor vehicle air conditioning systems, on the grounds that using flammable hydrocarbons in systems originally designed to carry non-flammable refrigerant presents a significant risk of fire or explosion.

Vendors and advocates of hydrocarbon refrigerants argue against such bans on the grounds that there have been very few such incidents relative to the number of vehicle air conditioning systems filled with hydrocarbons. One particular test, conducted by a professor at the University of New South Wales, unintentionally tested the worst-case scenario of a sudden and complete refrigerant expulsion into the passenger compartment followed by subsequent ignition. He and several others in the car sustained minor burns to their face, ears, and hands, and several observers received lacerations from the burst glass of the front passenger window. No one was seriously injured.

Global Production

Global LPG production reached over 292 million metric tons/yr in 2015, while global LPG consumption to over 284 mn t/yr. 62% of LPG is extracted from natural gas while the rest is produced by the petrochemical refineries from crude oil. 44% of global consumption is in the domestic sector. The USA is the leading producer and exporter of LPG.

Security of Supply

Because of the natural gas and the oil-refining industry, Europe is almost self-sufficient in LPG. Europe's security of supply is further safeguarded by:

- A wide range of sources, both inside and outside Europe;

- A flexible supply chain via water, rail and road with numerous routes and entry points into Europe;

According to 2010–12 estimates, proven world reserves of natural gas, from which most LPG is derived, stand at 300 trillion cubic meters (10,600 trillion cubic feet). Added to the LPG derived from cracking crude oil, this amounts to a major energy source that is virtually untapped and has massive potential. Production continues to grow at an average annual rate of 2.2%, virtually assuring that there is no risk of demand outstripping supply in the foreseeable future.

Comparison with Natural Gas

LPG is composed mainly of propane and butane, while natural gas is composed of the lighter methane and ethane. LPG, vaporised and at atmospheric pressure, has a higher calorific value (46 MJ/m^3 equivalent to 12.8 kWh/m^3) than natural gas (methane) (38 MJ/m^3 equivalent to 10.6 kWh/m^3), which means that LPG cannot simply be substituted for natural gas. In order to allow the use of the same burner controls and to provide for similar combustion characteristics, LPG can be mixed with air to produce a synthetic natural gas (SNG) that can be easily substituted. LPG/air mixing ratios average 60/40, though this is widely variable based on the gases making up the LPG. The method for determining the mixing ratios is by calculating the Wobbe index of the mix. Gases having the same Wobbe index are held to be interchangeable.

LPG-based SNG is used in emergency backup systems for many public, industrial and military installations, and many utilities use LPG peak shaving plants in times of high demand to make up shortages in natural gas supplied to their distributions systems. LPG-SNG installations are also used during initial gas system introductions, when the distribution infrastructure is in place before gas supplies can be connected. Developing markets in India and China (among others) use LPG-SNG systems to build up customer bases prior to expanding existing natural gas systems.

LPG-based SNG or natural gas with localized storage and piping distribution network to the house holds for catering to each cluster of 5000 domestic consumers can be planned under initial phase of city gas network system. This would eliminate the last mile LPG cylinders road transport which is a cause of traffic and safety hurdles in Indian cities. These localized natural gas networks are successfully operating in Japan with feasibility to get connected to wider networks in both villages and cities.

Environmental Effects

Commercially available LPG is currently derived mainly from fossil fuels. Burning LPG releases carbon dioxide, a greenhouse gas. The reaction also produces some carbon monoxide. LPG does, however, release less CO_2 per unit of energy than does coal or oil. It emits 81% of the CO_2 per kWh produced by oil, 70% of that of coal, and less than 50% of that emitted by coal-generated electricity distributed via the grid. Being a mix of propane and butane, LPG emits less carbon per joule than butane but more carbon per joule than propane.

LPG burns more cleanly than higher molecular weight hydrocarbons because it releases fewer particulates.

Fire/Explosion Risk and Mitigation

In a refinery or gas plant, LPG must be stored in pressure vessels. These containers are either cylindrical and horizontal or spherical. Typically, these vessels are designed and manufactured according to some code. In the United States, this code is governed by the American Society of Mechanical Engineers (ASME).

A spherical gas container typically found in refineries.

LPG containers have pressure relief valves, such that when subjected to exterior heating sources, they will vent LPGs to the atmosphere or a flare stack.

If a tank is subjected to a fire of sufficient duration and intensity, it can undergo a boiling liquid expanding vapor explosion (BLEVE). This is typically a concern for large refineries and petrochemical plants that maintain very large containers. In general, tanks are designed so that the product will vent faster than pressure can build to dangerous levels.

One remedy that is utilized in industrial settings is to equip such containers with a measure to provide a fire-resistance rating. Large, spherical LPG containers may have up to a 15 cm steel wall thickness. They are equipped with an approved pressure relief valve. A large fire in the vicinity of the vessel will increase its temperature and pressure. The relief valve on the top is designed to vent off excess pressure in order to prevent the rupture of the container itself. Given a fire of sufficient duration and intensity, the pressure being generated by the boiling and expanding gas can exceed the ability of the valve to vent the excess. If that occurs, an overexposed container may rupture violently, launching pieces at high velocity, while the released products can ignite as well, potentially causing catastrophic damage to anything nearby, including other containers.

People can be exposed to LPG in the workplace by breathing it in, skin contact, and eye contact. The Occupational Safety and Health Administration (OSHA) has set the legal limit (Permissible exposure limit) for LPG exposure in the workplace as 1000 ppm (1800 mg/m³) over an 8-hour workday. The National Institute for Occupational Safety and Health (NIOSH) has set a recommended exposure limit (REL) of 1000 ppm (1800 mg/m³) over an 8-hour workday. At levels of 2000 ppm, 10% of the lower explosive limit, LPG is considered immediately dangerous to life and health (due solely to safety considerations pertaining to risk of explosion).

Petroleum Coke

Petroleum coke, abbreviated coke or petcoke, is a final carbon-rich solid material that derives from oil refining, and is one type of the group of fuels referred to as cokes. Petcoke is the coke that, in particular, derives from a final cracking process—a thermo-based chemical engineering process that splits long chain hydrocarbons of petroleum into shorter chains—that takes place in units termed coker units. (Other types of coke are derived from coal). Stated succinctly, coke is the "carbonization product of high-boiling hydrocarbon fractions obtained in petroleum processing (heavy residues)." Petcoke is also produced in the production of synthetic crude oil (syncrude) from bitumen extracted from Canada's oil sands and from Venezuela's Orinoco oil sands.

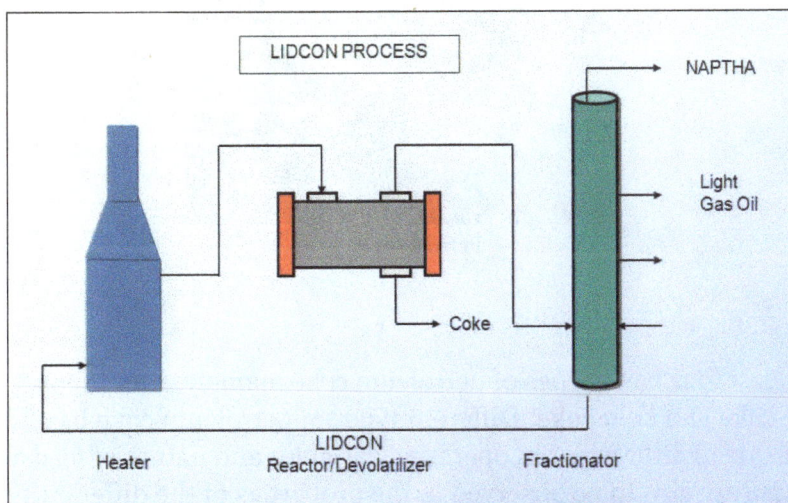

The above figure shows a schematic flow diagram of such a unit, where *residual oil* enters the process at the lower left, proceeds via pumps to the *main fractionator* (tall column at right), the residue of which, shown in green, is pumped via a furnace into the *coke drums* (two columns left and center) where the final carbonization takes place, at high temperature and pressure, in the presence of steam.

In petroleum coker units, residual oils from other distillation processes used in petroleum refining are treated at a high temperature and pressure leaving the petcoke after driving off gases and volatiles, and separating off remaining light and heavy oils. These processes are termed "coking processes", and most typically employ chemical engineering plant operations for the specific process of delayed coking.

This coke can either be fuel grade (high in sulfur and metals) or anode grade (low in sulfur and metals). The raw coke directly out of the coker is often referred to as green coke. In this context, "green" means unprocessed. The further processing of green coke by calcining in a rotary kiln removes residual volatile hydrocarbons from the coke. The calcined petroleum coke can be further processed in an anode baking oven to produce anode coke of the desired shape and physical properties. The anodes are mainly used in the aluminium and steel industry.

Petcoke is over 80% carbon and emits 5% to 10% more carbon dioxide (CO_2) than coal on a per-unit-of-energy basis when it is burned. As petcoke has a higher energy content, petcoke emits between 30 and 80 percent more CO_2 than coal per unit of weight. The difference between coal and coke in CO_2 production per unit of energy produced depends upon the moisture in the coal, which increases the CO_2 per unit of energy – heat of combustion – and on the volatile hydrocarbons in coal and coke, which decrease the CO_2 per unit of energy.

Petroleum coke

Types

There are at least four basic types of petroleum coke, namely, needle coke, honeycomb coke, sponge coke and shot coke. Different types of petroleum coke have different microstructures due to differences in operating variables and nature of feedstock. Significant differences are also to be observed in the properties of the different types of coke, particularly ash and volatile matter contents.

Needle coke, also called acicular coke, is a highly crystalline petroleum coke used in the production of electrodes for the steel and aluminium industries and is particularly valuable because the electrodes must be replaced regularly. Needle coke is produced exclusively from either FCC decant oil or coal tar pitch.

Honeycomb coke is an intermediate coke, with ellipsoidal pores that are uniformly distributed. Compared to needle coke, honeycomb coke has a lower coefficient of thermal expansion and a lower electrical conductivity.

Composition

Petcoke, altered through the process of calcined which it is heated or refined raw coke eliminates much of the component of the resource. Usually petcoke when refined does not release the heavy metals as volatiles or emissions.

Depending on the petroleum feed stock used, the composition of petcoke may vary but the main thing is that it is primarily carbon. Petcoke is primarily made up of carbon, when in pure form petcoke can weigh 98-99% which creates a carbon based compound with the hydrogen filling in. In raw form hydrogen can have a weight range of 3.0- 4.0%. Petcoke in its raw (green coke) nitrogen at 0.1- 0.5% and sulfur 0.2- 6.0% become emissions after the coke calcined.

Through process of thermal processing the composition in weight is reduced with the volatile matter and sulfur being emitted. This process ends in the honeycomb petcoke which according to the name giving is a solid carbon structure with holes in it.

Depending on the petroleum feed stock used, the composition of petcoke may vary but the main thing is that it is primarily carbon. Petcoke is primarily made up of carbon, when in pure form petcoke can weigh 98-99% which creates a carbon based compound with the hydrogen filling in. In raw form hydrogen can have a weight range of 3.0-4.0%. Petcoke in its raw(green coke) nitrogen at 0.1- 0.5% and sulfur 0.2- 6.0% become emissions after the coke calcined.

Other heavy metals found can be found with in petcoke as impurities due to that some of these metals come in after processing as volatile.

Fuel-grade

Fuel-grade coke is classified as either sponge coke or shot coke morphology. While oil refiners have been producing coke for over 100 years, the mechanisms that cause sponge coke or shot coke to form are not well understood and cannot be accurately predicted. In general, lower temperatures and higher pressures promote sponge coke formation. Additionally, the amount of heptane insolubles present and the fraction of light components in the coker feed contribute.

While its high heat and low ash content make it a decent fuel for power generation in coal-fired boilers, petroleum coke is high in sulfur and low in volatile content, and this poses environmental (and technical) problems with its combustion. Its gross calorific value (HHV) is nearly 8000 Kcal/kg which is twice the value of average coal used in electricity generation. A common choice of sulfur recovering unit for burning petroleum coke is the SNOX Flue gas desulfurisation technology, which is based on the well-known WSA Process. Fluidized bed combustion is commonly used to burn petroleum coke. Gasification is increasingly used with this feedstock (often using gasifiers placed in the refineries themselves).

Calcined

Calcined petroleum coke (CPC) is the product from calcining petroleum coke. This coke is the product of the coker unit in a crude oil refinery. The calcined petroleum coke is used to make anodes for the aluminium, steel and titanium smelting industry. The green coke must have sufficiently low metal content to be used as anode material. Green coke with this low metal content is called anode-grade coke. When green coke has excessive metal content, it is not calcined and is used as fuel-grade coke in furnaces.

Desulfurization

A high sulfur content in petcoke reduces its market value, and may preclude its use as fuel due to restrictions on sulfur oxides emissions for environmental reasons. Methods have thus been proposed to reduce or eliminate the sulfur content of petcoke. Most of them involve the desorption of the inorganic sulfur present in the pores or surface of the coke, and the partition and removal of the organic sulfur attached to the aromatic carbon skeleton.

Potential petroleum desulfurization techniques can be classified as follows:

1. Solvent extraction.

2. Chemical treatment.

3. Thermal desulfurization.

4. Desulfurization in an oxidizing atmosphere.

5. Desulfurization in an atmosphere of sulfur-bearing gas.

6. Desulfurization in an atmosphere of hydrocarbon gases.

7. Hydrodesulfurization.

As of 2011 there was no commercial process available to desulfurize petcoke.

Storage, Disposal and Sale

Nearly pure carbon, petcoke is a potent source of carbon dioxide if burned.

Petroleum coke may be stored in a pile near an oil refinery pending sale. For example, in 2013 a large stockpile owned by Koch Carbon near the Detroit River was produced by a Marathon Petroleum refinery in Detroit which had begun refining bitumen from the oil sands of Alberta in November 2012. Large stockpiles of petcoke also existed in Canada as of 2013, and China and Mexico were markets for petcoke exported from California to be used as fuel. As of 2013 Oxbow Corporation, owned by William I. Koch, was a major dealer in petcoke, selling 11 million tons annually.

In 2017 a quarter of US exports of the fuel went to India, an Associated Press investigation found. In 2016 this amounted to more than eight million metric tons, more than 20 times as much as in 2010. India's Environmental Pollution Control Authority tested imported petcoke in use near New Delhi, and found sulfur levels 17 times the legal limit.

The International Convention for Prevention of Pollution from Ships (MARPOL 73/78), adopted by the IMO, has mandated that marine vessels shall not consume residual fuel oils (bunker fuel, etc) with a sulfur content greater than 0.1% from the year 2020. Nearly 38% of residual fuel oils are consumed in the shipping sector. In the process of converting excess residual oils into lighter oils by coking processes, pet coke is generated as a byproduct. Pet coke availability is expected to increase in the future due to falling demand for residual oil. Pet coke is also used in methanation plants to produce synthetic natural gas, etc. in order to avoid a pet coke disposal problem.

Health Hazards

Petroleum coke is sometimes a source of fine dust, which can penetrate the filtering process of the human airway, lodge in the lungs and cause serious health problems. Studies have shown that petroleum coke itself has a low level of toxicity and there is no evidence of carcinogenicity.

Petroleum coke can contain vanadium, a toxic metal. Vanadium was found in the dust collected in occupied dwellings near the petroleum coke stored next to the Detroit River. Vanadium is toxic in tiny quantities, 0.8 micrograms per cubic meter of air, according to the EPA.

According to multiple EPA studies and analyses, petroleum coke has a low health hazard potential in humans. It does not have any observable carcinogenic, developmental, or reproductive effects. During animal case studies repeated-dose chronic inhalation did show respiratory inflammation due to dust particles, but not specific to petroleum coke.

Environmental Hazards

Environmental concerns stem from the storage and combustion of petcoke. By-waste accumulates as petcoke is processed, making waste management an issue. Petcoke's high silt content of 21.2% increases the risk of fugitive dust drifting away from petcoke mounds under heavy wind. An estimated 100 tons of petcoke fugitive dust including PM10 and PM2.5 are released into the atmosphere per year in the United States. Waste management and release of fugitive dust is especially an issue in the cities of Chicago, Detroit and Green bay.

Externalities stem from petcoke that cause potential environmental impacts. Petcoke is composed of 90% elemental carbon by weight which is converted to CO_2 during combustion. Use of petcoke also produces emissions of sulfur, and the potential for water pollution through nickel and vanadium runoff from refining and storage.

Crude Oil

Crude oil is a liquid petroleum that is found accumulated in various porous rock formations in Earth's crust and is extracted for burning as fuel or for processing into chemical products.

Crude oil is a mixture of comparatively volatile liquid hydrocarbons (compounds composed mainly of hydrogen and carbon), though it also contains some nitrogen, sulfur, and oxygen. Those elements form a large variety of complex molecular structures, some of which cannot be readily identified. Regardless of variations, however, almost all crude oil ranges from 82 to 87 percent carbon by weight and 12 to 15 percent hydrogen by weight.

Crude oils are customarily characterized by the type of hydrocarbon compound that is most prevalent in them: paraffins, naphthenes, and aromatics. Paraffins are the most common hydrocarbons in crude oil; certain liquid paraffins are the major constituents of gasoline (petrol) and are therefore highly valued. Naphthenes are an important part of all liquid refinery products, but they also form some of the heavy asphaltlike residues of refinery processes. Aromatics generally constitute only a small percentage of most crudes. The most common aromatic in crude oil is benzene, a popular building block in the petrochemical industry.

Because crude oil is a mixture of such widely varying constituents and proportions, its physical properties also vary widely. In appearance, for instance, it ranges from colourless to black. Possibly the most important physical property is specific gravity (i.e., the ratio of the weight of equal volumes of a crude oil and pure water at standard conditions). In laboratory measurement of specific gravity, it is customary to assign pure water a measurement of 1; substances lighter than water, such as crude oil, would receive measurements less than 1. The petroleum industry, however, uses the American Petroleum Institute (API) gravity scale, in which pure water has been arbitrarily assigned an API gravity of 10°. Liquids lighter than water, such as oil, have API gravities numerically greater than 10. On the basis of their API gravity, crude oils can be classified as heavy, medium, and light as follows:

- Heavy: 10–20° API gravity.

- Medium: 20–25° API gravity.

- Light: above 25° API gravity.

Crude oil also is categorized as "sweet" or "sour" depending on the level of sulfur, which occurs either as elemental sulfur or in compounds such as hydrogen sulfide. Sweet crudes have sulfur contents of 0.5 percent or less by weight, and sour crudes have sulfur contents of 1 percent or more by weight. Generally, the heavier the crude oil, the greater its sulfur content. Excess sulfur is removed from crude oil during refining, because sulfur oxides released into the atmosphere during combustion of oil are a major pollutant.

In the United States, the conventional practice for the petroleum industry is to measure capacity by volume and to use the English system of measurement. For this reason, crude oil in the United States is measured in barrels, each barrel containing 42 gallons of oil. Most other areas of the world define capacity by the weight of materials processed and record measurements in metric units; therefore, crude oil outside the United States is usually measured in metric tons. A barrel of API 30° light oil would weigh about 139 kg (306 pounds). Conversely, a metric ton of API 30° light oil would be equal to approximately 252 imperial gallons, or about 7.2 U.S. barrels.

Crude oil occurs underground, at various pressures depending on depth. It can contain considerable natural gas, kept in solution by the pressure. In addition, water often flows into an oil well along with liquid crude and gas. All these fluids are collected by surface equipment for separation. Clean crude oil is sent to storage at near atmospheric pressure, usually aboveground in cylindrical steel tanks that may be as large as 30 metres (100 feet) in diameter and 10 metres (33 feet) tall. Often crude oil must be transported from widely distributed production sites to treatment plants and refineries. Overland movement is largely through pipelines. Crude from more isolated wells is collected in tank trucks and taken to pipeline terminals; there is also some transport in specially constructed railroad cars. Overseas transport is conducted in specially designed tanker ships. Tanker capacities vary from less than 100,000 barrels to more than 3,000,000 barrels.

The primary destination of crude oil is a refinery. There any combination of three basic functions is carried out: (1) separating the many types of hydrocarbon present in crude oils into fractions of more closely related properties, (2) chemically converting the separated hydrocarbons into more desirable reaction products, and (3) purifying the products of unwanted elements and compounds. The main process for separating the hydrocarbon components of crude oil is fractional distillation. Crude oil fractions separated by distillation are passed on for subsequent processing into numerous products, ranging from gasoline and diesel fuel to heating oil to asphalt. Given the pattern of modern demand (which tends to be highest for transportation fuels such as gasoline), the market value of a crude oil generally rises with increasing yields of light products.

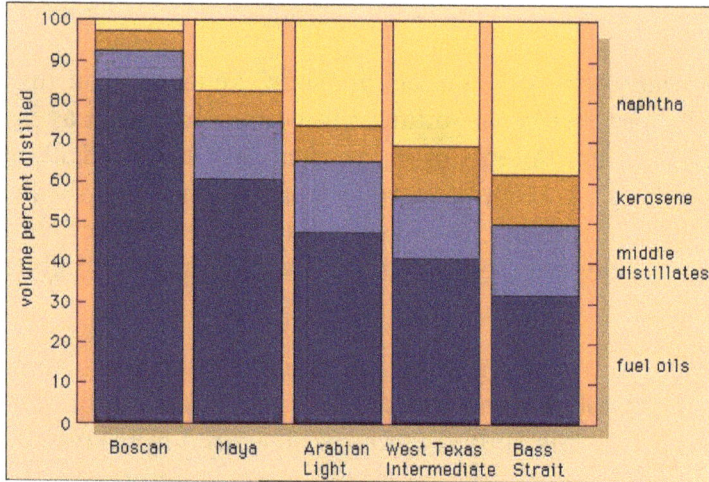

Product content of five major crude oils.

Diesel Fuel

Diesel fuel, also called diesel oil is a combustible liquid used as fuel for diesel engines, ordinarily obtained from fractions of crude oil that are less volatile than the fractions used in gasoline. In diesel engines the fuel is ignited not by a spark, as in gasoline engines, but by the heat of air compressed in the cylinder, with the fuel injected in a spray into the hot compressed air. Diesel fuel releases more energy on combustion than equal volumes of gasoline, so diesel engines generally produce better fuel economy than gasoline engines. In addition, the production of diesel fuel requires fewer refining steps than gasoline, so retail prices of diesel fuel traditionally have been lower than those of gasoline (depending on the location, season, and taxes and regulations). On the other hand, diesel fuel, at least as traditionally formulated, produces greater quantities of certain air pollutants such as sulfur and solid carbon particulates, and the extra refining steps and emission-control mechanisms put into place to reduce those emissions can act to reduce the price advantages of diesel over gasoline. In addition, diesel fuel emits more carbon dioxide per unit than gasoline, offsetting some of its efficiency benefits with its greenhouse gas emissions.

Several grades of diesel fuel are manufactured—for example, "light-middle" and "middle" distillates for high-speed engines with frequent and wide variations in load and speed (such as trucks and automobiles) and "heavy" distillates for low- and medium-speed engines with sustained loads and speeds (such as trains, ships, and stationary engines). Performance criteria are cetane number (a measure of ease of ignition), ease of volatilization, and sulfur content. The highest grades, for automobile and truck engines, are the most volatile, and the lowest grades, for low-speed engines, are the least volatile, leave the most carbon residue, and commonly have the highest sulfur content.

Sulfur is a critical polluting component of diesel and has been the object of much regulation. Traditional "regular" grades of diesel fuel contained as much as 5,000 parts per million (ppm) by weight sulfur. In the 1990s "low sulfur" grades containing no more than 500 ppm sulfur were introduced, and in the following years even lower levels of sulfur were required. Regulations in the United States required that by 2010 diesel fuels sold for highway vehicles be "ultra-low sulfur" (ULSD) grades, containing a maximum of 15 ppm. In the European Union, regulations required that from 2009 diesel fuel sold for road vehicles be only so-called "zero-sulfur," or "sulfur-free," diesels, containing no more than 10 ppm. Lower sulfur content reduces emissions of sulfur compounds implicated in acid rain and allows diesel vehicles to be equipped with highly effective emission-control systems that would otherwise be damaged by higher concentrations of sulfur. Heavier grades of diesel fuel, made for use by off-road vehicles, ships and boats, and stationary engines, are generally allowed higher sulfur content, though the trend has been to reduce limits in those grades as well.

In addition to traditional diesel fuel refined from petroleum, it is possible to produce so-called synthetic diesel, or Fischer-Tropsch diesel, from natural gas, from synthesis gas derived from coal, or from biogas obtained from biomass. Also, biodiesel, a bio-fuel, can be made primarily from oily plants such as the soybean or oil palm. These alternative diesel fuels can be blended with traditional diesel fuel or used alone in diesel engines without modification, and they have very low sulfur content. Alternative diesel fuels are often proposed as means to reduce dependence on petroleum and to reduce overall emissions, though only biodiesel can provide a life cycle carbon dioxide benefit.

Petroleum Ether

Petroleum ether is the petroleum fraction consisting of aliphatic hydrocarbons and boiling in the range 35–60 °C; commonly used as a laboratory solvent. Despite the name, petroleum ether is not classified as an ether; the term is used only figuratively, signifying extreme lightness and volatility.

Properties

The very lightest, most volatile liquid hydrocarbon solvents that can be bought from laboratory chemical suppliers may also be offered under the name petroleum ether. Petroleum ether consists mainly of aliphatic hydrocarbons and is usually low on aromatics. It is commonly hydrodesulfurized and may be hydrogenated to reduce the amount of aromatic and other unsaturated hydrocarbons. Petroleum ether bears normally a descriptive suffix giving the boiling range. Thus, from the leading international laboratory chemical suppliers it is possible to buy various petroleum ethers with boiling ranges suchs 100-140 °C may be called petroleum ether, rather than petroleum spirit.

It is not advisable to employ a fraction with a wider boiling point range than 20 °C, because of possible loss of the more volatile portion during its use in recrystallisation, etc. and consequent different solubility relations of the higher boiling residue.

Most of the unsaturated hydrocarbons may be removed by shaking two or three times with 10% of the volume of concentrated sulfuric acid; vigorous shaking is then continued with successive portions of a concentrated solution of potassium permanganate in 10% sulfuric acid until the color of the permanganate remains unchanged. The solvent is then thoroughly washed with sodium carbonate solution and then with water, dried over anhydrous calcium chloride, and distilled. If required perfectly dry, it can be allowed to stand over sodium wire, or calcium hydride.

Safety

Petroleum ethers are extremely volatile, have very low flash points, and present a significant fire hazard. Fires should be fought with foam, carbon dioxide, dry chemical or carbon tetrachloride.

The naphtha mixtures that are distilled at a lower boiling temperature have a higher volatility and, generally speaking, a higher degree of toxicity than the higher boiling fractions.

Exposure to petroleum ether occurs most commonly by either inhalation or through skin contact. Petroleum ether is metabolized by the liver with a biological half-life of 46–48 h.

Inhalation overexposure causes primarily central nervous system (CNS) effects (headaches, dizzines, nausea, fatigue, and incoordination). In general, the toxicity is more pronounced with petroleum ethers containing higher concentrations of aromatic compounds. n-Hexane is known to cause axonal damage in peripheral nerves.

Skin contact can cause allergic contact dermatitis.

Oral ingestion of hydrocarbons often is associated with symptoms of mucous membrane irritation, vomiting, and central nervous system depression. Cyanosis, tachycardia, and tachypnea may appear as a result of aspiration, with subsequent development of chemical pneumonitis. Other clinical findings include albuminuria, hematuria, hepatic enzyme derangement, and cardiac arrhythmias. Doses as low as 10 ml orally have been reported to be potentially fatal, whereas some patients have survived the ingestion of 60 ml of petroleum distillates. A history of coughing or choking in association with vomiting strongly suggests aspiration and hydrocarbon pneumonia. Hydrocarbon pneumonia is an acute hemorrhagic necrotizing disease that can develop within 24 h after the ingestion. Pneumonia may require several weeks for complete resolution.

Intravenous administration produces fever and local tissue damage.

Petroleum-derived distillates have not been shown to be carcinogenic in humans. Petroleum ether degrades rapidly in soil and water.

Petroleum Jelly

Petroleum jelly, petrolatum, white petrolatum, soft paraffin, or multi-hydrocarbon, is a semi-solid mixture of hydrocarbons (with carbon numbersmainly higher than 25), originally promoted as a topical ointment for its healing properties.

After petroleum jelly became a medicine chest staple, consumers began to use it for many ailments, as well as cosmetic purposes, including toenail fungus, genital rashes (non-STD), nosebleeds, diaper rash, and chest colds. Its folkloric medicinal value as a "cure-all" has since been limited by better scientific understanding of appropriate and inappropriate uses. It is recognized by the U.S. Food and Drug Administration (FDA) as an approved over-the-counter (OTC) skin protectant and remains widely used in cosmeticskin care.

Petroleum jelly.

The raw material for petroleum jelly was discovered in 1859 in Titusville, Pennsylvania, United States of America, on some of the country's first oil rigs. Workers disliked the paraffin-like material forming on rigs because it caused them to malfunction, but they used it on cuts and burns because they believed that it hastened healing.

Robert Chesebrough, a young chemist whose previous work of distilling fuel from the oil of sperm whales had been rendered obsolete by petroleum, went to Titusville to see what new materials had commercial potential. Chesebrough took the unrefined black "rod wax", as the drillers called it, back to his laboratory to refine it and explore potential uses. Chesebrough discovered that by distilling the lighter, thinner oil products from the rod wax, he could create a light-colored gel. Chesebrough patented the process of making petroleum jelly by U.S. Patent 127,568 in 1872. The process involved vacuum distillation of the crude material followed by filtration of the still residue through bone char.

Chesebrough traveled around New York demonstrating the product to encourage sales by burning his skin with acid or an open flame, then spreading the ointment on his injuries and showing his past injuries healed, he claimed, by his miracle product.

He opened his first factory in 1870 in Brooklyn using the name Vaseline.

Physical Properties

Petroleum jelly is a mixture of hydrocarbons, with a melting point that depends on the exact proportions. The melting point is typically between 40 and 70 °C (105 and 160 °F). It is flammable only when heated to liquid; then the fumes will light, not the liquid itself, so a wick material like leaves, bark, or small twigs is needed to ignite petroleum jelly. It is colorless or has a pale yellow color (when not highly distilled), translucent, and devoid of taste and smell when pure. It does not oxidize on exposure to the air and is not readily acted on by chemical reagents. It is insoluble in water. It is soluble in dichloromethane, chloroform, benzene, diethyl ether, carbon disulfide and oil of turpentine.

Depending on the specific application of petroleum jelly, it may be USP, B.P., or Ph. Eur. grade. This pertains to the processing and handling of the petroleum jelly so it is suitable for medicinal and personal-care applications.

Comparison with Glycerol

Because they feel similar when applied to human skin, there is a common misconception that petroleum jelly and glycerol (glycerine) are physically similar. Petroleum jelly is a non-polar hydrophobic (water-repelling) hydrocarbon and insoluble in water. Glycerol is an alcohol that is strongly hydrophilic (water-attracting): by continuously absorbing moisture from the air (hygroscopic), it produces the feeling of wetness on the skin. This feeling of wetness is similar to the feeling of greasiness produced by petroleum jelly.

Uses

Most uses of petroleum jelly exploit its lubricating and coating properties.

Medical Treatment

Vaseline brand First Aid Petroleum Jelly, or carbolated petroleum jelly containing phenol to give the jelly additional antibacterial effect, has been discontinued. During World War II, a variety of petroleum jelly called *red veterinary petrolatum,* or Red Vet Pet for short, was often included in life raft survival kits. Acting as a sunscreen, it provides protection against ultraviolet rays.

The American Academy of Dermatology recommends keeping skin injuries moist with petroleum jelly to reduce scarring. A verified medicinal use is to protect and prevent moisture loss of the skin of a patient in the initial post-operative period following laser skin resurfacing.

There is one case report published in 1994 indicating petroleum jelly should not be applied to the inside of the nose due to the risk of lipid pneumonia, but this was only ever reported in one patient. However, petroleum jelly is used extensively by otolaryngologists—ear, nose, and throat surgeons—for nasal moisture and epistaxis treatment, and to combat nasal crusting. Large studies have found petroleum jelly applied to the nose for short durations to have no significant side effects.

Historically, it was also consumed for internal use and even promoted as "Vaseline confection".

Skin and Hair Care

Most petroleum jelly today is used as an ingredient in skin lotions and cosmetics, providing various types of skin care and protection by minimizing friction or reducing moisture loss, or by functioning as a grooming aid, e. g. pomade.

Preventing Moisture Loss

By reducing moisture loss, petroleum jelly can prevent chapped hands and lips, and soften nail cuticles.

This property is exploited to provide heat insulation: petroleum jelly can be used to keep swimmers warm in water when training or during channel crossings or long ocean swims. It can prevent chilling of the face due to evaporation of skin moisture during cold weather outdoor sports.

Hair Grooming

In the first part of the twentieth century, petroleum jelly, either pure or as an ingredient, was also popular as a hair pomade. When used in a 50/50 mixture with pure beeswax, it makes an effective moustache wax.

Skin Lubrication

Petroleum jelly can be used to reduce the friction between skin and clothing during various sport activities, for example to prevent chafing of the seat region of cyclists or the nipples of long distance runners wearing loose T-shirts, and is commonly used in the crotch area of wrestlers and footballers.

Petroleum jelly is commonly used as a personal lubricant because it does not dry out like water-based lubricants, and has a distinctive "feel", different from that of K-Y and related methylcellulose products. However, it is not recommended for use with condoms during sexual activity because it swells latex and thus increases the chance of rupture. It is also not recommended for vaginal intercourse because it may increase the risk of yeast infection and bacterial vaginosis in women.

Product Care and Protection

Coating

Petroleum jelly can be used to coat corrosion-prone items such as metallic trinkets, non-stainless steel blades, and gun barrels prior to storage as it serves as an excellent and inexpensive water repellent. It is used as an environmentally friendly underwater antifouling coating for motor boats and sailing yachts. It was recommended in the Porsche owner's manual as a preservative for light alloy (alleny) anodized Fuchs wheels to protect them against corrosion from road salts and brake dust. "Every three months (after regular cleaning) the wheels should be coated with petroleum jelly."

Finishing

It can be used to finish and protect wood, much like a mineral oil finish. It is used to condition and protect smooth leather products like bicycle saddles, boots, motorcycle clothing, and used to put a shine on patent leather shoes (when applied in a thin coat and then gently buffed off).

Lubrication

Petroleum jelly can be used to lubricate zippers and slide rules. It was also recommended by Porsche in maintenance training documentation for lubrication (after cleaning) of "Weatherstrips on Doors, Hood, Tailgate, Sun Roof". The publication states "before applying a new coat of lubricant" "Only acid-free lubricants may be used, for example: glycerine, Vaseline, tire mounting paste, etc. These lubricants should be rubbed in, and excessive lubricant wiped off with a soft cloth." It is used in bullet lubricant compounds.

Industrial Production Processes

Petroleum jelly is a useful material when incorporated into candle wax formulas. The petroleum jelly softens the overall blend, allows the candle to incorporate additional fragrance oil, and facilitates adhesion to the sidewall of the glass. Petroleum jelly is used to moisten nondrying modelling clay such as plasticine, as part of a mix of hydrocarbons including those with greater (paraffin wax) and lesser (mineral oil) molecular weights. It is used as a tack reducer additive to printing inks to reduce paper lint "picking" from uncalendered paper stocks. It can be used as a release agent for plaster molds and castings. It is used in the leather industry as a waterproofing cream.

Other

Explosives

Petroleum jelly is mixed with a high proportion of strong inorganic chlorates due to it acting as a plasticizer and a fuel source. An example of this is Cheddite C which consists of a ratio of 9:1, $KClO_3$ to petroleum jelly. This mixture is unable to detonate without the use of a blasting cap. It is also used as a stabiliser in the manufacture of the propellant Cordite.

Mechanical and Barrier Functions

Petroleum jelly can be used to coat the inner walls of terrariums to prevent animals crawling out and escaping.

A stripe of petroleum jelly can be used to prevent the spread of a liquid. For example, it can be applied close to the hairline when using a home hair dye kit to prevent the hair dye from irritating or staining the skin. It is also used to prevent diaper rash.

Surface Cleansing

Petroleum jelly is used to gently clean a variety of surfaces, ranging from makeup removal from faces to tar stain removal from leather.

Pet care

Petroleum jelly is used to moisturize the paws of dogs, and to inhibit fungal growth on aquatic turtles' shells. It is a common ingredient in hairball remedies for domestic cats.

Clean-up

Petroleum jelly is very sticky and hard to remove from non-biological surfaces with the usual and customary cleaning agents typically found in the home. It may be dissolved with paint thinner or other petroleum solvents such as acetone, which dissolves many plastics.

Petroleum jelly is slightly soluble in alcohol. To avoid damage to plastics and minimize ventilation issues, isopropyl (rubbing) alcohol can be used to remove petroleum jelly from most surfaces. Isopropyl alcohol is inert to most household surfaces, including almost every plastic, and removes petroleum jelly efficiently. While alcohol causes fewer ventilation problems than petroleum solvents, ventilation is still recommended, especially if large surface areas are involved.

Petroleum jelly is also soluble in lower-molecular-weight oils. Using an oil to dissolve the petroleum jelly first can render it more soluble to solvents and soaps that would not dissolve pure petroleum jelly. Vegetable oils such as canola and olive oil are commonly used to aid in the removal of petroleum jelly from hair and skin.

Health

In 2015, German consumer watchdog Stiftung Warentest analyzed cosmetics containing mineral oils. After developing a new detection method they found high concentrations of Mineral Oil Aromatic Hydrocarbons (MOHA) and even polyaromatics in products containing mineral oils with Vaseline products containing the most MOHA of all tested cosmetics (up to 9%). The European Food Safety Authority sees MOHA and polyaromatics as possibly carcinogenic. Based on the results, Stiftung Warentest warns not to use Vaseline or any product that is based on mineral oils for lip care.

Gasoline

Gasoline, or petrol, is a colorless petroleum-derived flammable liquid that is used primarily as a fuel in spark-ignited internal combustion engines. It consists mostly of organic compounds obtained by the fractional distillation of petroleum, enhanced with a variety of additives. On average, a 42-U.S.-gallon (160-liter) barrel of crude oil yields about 19 U.S. gallons (72 liters) of gasoline (among other refined products) after processing in an oil refinery, though this varies based on the crude oil assay.

The characteristic of a particular gasoline blend to resist igniting too early (which causes knocking and reduces efficiency in reciprocating engines) is measured by its octane rating which is produced in several grades. Tetraethyl lead and other lead compounds are no longer used in most areas to increase octane rating (still used in aviation and auto-racing). Other chemicals are frequently added to gasoline to improve chemical stability and performance characteristics, control corrosiveness and provide fuel system cleaning. Gasoline may contain oxygen-containing chemicals such as ethanol, MTBE or ETBE to improve combustion.

A typical gasoline container.

Gasoline used in internal combustion engines can have significant effects on the local environment, and is also a contributor to global human carbon dioxide emissions. Gasoline can also enter the environment uncombusted, both as liquid and as vapor, from leakage and handling during production, transport and delivery (e.g., from storage tanks, from spills, etc.). As an example of efforts to control such leakage, many underground storage tanks are required to have extensive measures in place to detect and prevent such leaks. Gasoline contains benzene and other known carcinogens.

Chemical Analysis and Production

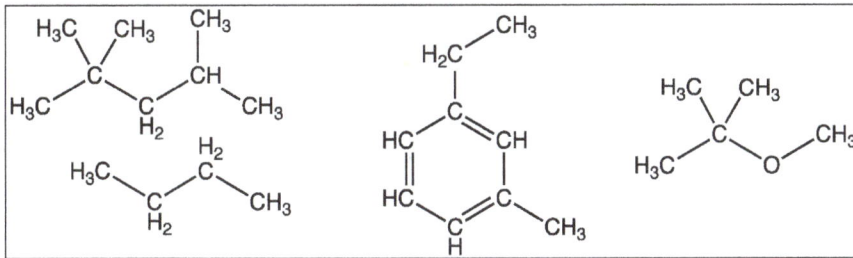

Some of the main components of gasoline: Isooctane, butane,
3-ethyltoluene, and the octane enhancer MTBE.

Gasoline is produced in oil refineries. Roughly 19 U.S. gallons (72 L) of gasoline is derived from a 42-U.S.-gallon (160 L) barrel of crude oil. Material separated from crude oil via distillation, called virgin or straight-run gasoline, does not meet specifications for modern engines, but can be pooled to the gasoline blend.

A pumpjack in the United States.

An oil rig in the Gulf of Mexico.

The bulk of a typical gasoline consists of a homogeneous mixture of small, relatively lightweight hydrocarbons with between 4 and 12 carbon atoms per molecule (commonly referred to as C4–C12). It is a mixture of paraffins (alkanes), olefins (alkenes) and cycloalkanes (naphthenes). The usage of the terms *paraffin* and *olefin* in place of the standard chemical nomenclature *alkane* and *alkene*, respectively, is particular to the oil industry. The actual ratio of molecules in any gasoline depends upon:

- The oil refinery that makes the gasoline, as not all refineries have the same set of processing units;

- The crude oil feed used by the refinery;

- The grade of gasoline (in particular, the octane rating).

The various refinery streams blended to make gasoline have different characteristics. Some important streams include:

- Straight-run gasoline, commonly referred to as *naphtha*, which is distilled directly from crude oil. Once the leading source of fuel, its low octane rating required lead additives. It is low in aromatics (depending on the grade of the crude oil stream) and contains some cycloalkanes (naphthenes) and no olefins (alkenes). Between 0 and 20 percent of this stream is pooled into the finished gasoline, because the supply of this fraction is insufficient and its RON is too low. The chemical properties (namely RON and Reid vapor pressure) of the straight-run gasoline can be improved through reforming and isomerisation. However, before feeding those units, the naphtha needs to be split into light and heavy naphtha. Straight-run gasoline can also be used as a feedstock for steam-crackers to produce olefins.

- Reformate, produced in a catalytic reformer, has a high octane rating with high aromatic content and relatively low olefin content. Most of the benzene, toluene and xylene (the so-called BTX hydrocarbons) are more valuable as chemical feedstocks and are thus removed to some extent.

- Catalytic cracked gasoline, or catalytic cracked naphtha, produced with a catalytic cracker, has a moderate octane rating, high olefin content and moderate aromatic content.

- Hydrocrackate (heavy, mid and light), produced with a hydrocracker, has a medium to low octane rating and moderate aromatic levels.

- Alkylate is produced in an alkylation unit, using isobutane and olefins as feedstocks. Finished alkylate contains no aromatics or olefins and has a high MON.

- Isomerate is obtained by isomerizing low-octane straight-run gasoline into iso-paraffins (non-chain alkanes, such as isooctane). Isomerate has a medium RON and MON, but no aromatics or olefins.

- Butane is usually blended in the gasoline pool, although the quantity of this stream is limited by the RVP specification.

The terms above are the jargon used in the oil industry and terminology varies.

Currently, many countries set limits on gasoline aromatics in general, benzene in particular, and olefin (alkene) content. Such regulations have led to an increasing preference for high-octane pure paraffin (alkane) components, such as alkylate, and are forcing refineries to add processing units to reduce benzene content. In the European Union, the benzene limit is set at 1% by volume for all grades of automotive gasoline.

Gasoline can also contain other organic compounds, such as organic ethers (deliberately added), plus small levels of contaminants, in particular organosulfur compounds (which are usually removed at the refinery).

Physical Properties

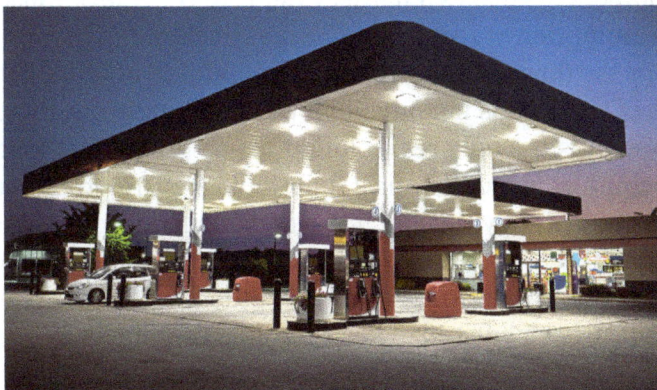

Gasoline station.

Density

The specific gravity of gasoline is from 0.71 to 0.77, with higher densities having a greater volume of aromatics. Finished marketable gasoline is traded (in Europe) with a standard reference of 0.755 kg/L (6.30 lb/US gal), and its price is escalated or de-escalated according to its actual density. Because of its low density, gasoline floats on water, and so water cannot generally be used to extinguish a gasoline fire unless applied in a fine mist.

Stability

Quality gasoline should be stable for six months if stored properly, but as gasoline is a mixture rather than a single compound, it will break down slowly over time due to the separation of the components. Gasoline stored for a year will most likely be able to be burned in an internal combustion engine without too much trouble but the effects of long-term storage will become more noticeable with each passing month until a time comes when the gasoline should be diluted with ever-increasing amounts of freshly made fuel so that the older gasoline may be used up. If left undiluted, improper operation will occur and this may include engine damage from misfiring or the lack of proper action of the fuel within a fuel injection system and from an onboard computer attempting to compensate (if applicable to the vehicle). Gasoline should ideally be stored in an airtight container (to prevent oxidation or water vapor mixing in with the gas) that can withstand the vapor pressure of the gasoline without venting (to prevent the loss of the more volatile fractions) at a stable cool temperature (to reduce the excess pressure from liquid expansion and to reduce the rate of any decomposition reactions). When gasoline is not stored correctly, gums and solids may result, which can corrode system components and accumulate on wetted surfaces, resulting in a condition called "stale fuel". Gasoline containing ethanol is especially subject to absorbing atmospheric moisture, then forming gums, solids or two phases (a hydrocarbon phase floating on top of a water-alcohol phase).

The presence of these degradation products in the fuel tank or fuel lines plus a carburetor or fuel injection components makes it harder to start the engine or causes reduced engine performance. On resumption of regular engine use, the buildup may or may not be eventually cleaned out by the flow of fresh gasoline. The addition of a fuel stabilizer to gasoline can extend the life of fuel that is not or cannot be stored properly, though removal of all fuel from a fuel system is the only real solution to the problem of long-term storage of an engine or a machine or vehicle. Typical fuel stabilizers are proprietary mixtures containing mineral spirits, isopropyl alcohol, 1,2,4-trimethylbenzene or other additives. Fuel stabilizers are commonly used for small engines, such as lawnmower and tractor engines, especially when their use is sporadic or seasonal (little to no use for one or more seasons of the year). Users have been advised to keep gasoline containers more than half full and properly capped to reduce air exposure, to avoid storage at high temperatures, to run an engine for ten minutes to circulate the stabilizer through all components prior to storage, and to run the engine at intervals to purge stale fuel from the carburetor.

Gasoline stability requirements are set by the standard ASTM D4814. This standard describes the various characteristics and requirements of automotive fuels for use over a wide range of operating conditions in ground vehicles equipped with spark-ignition engines.

Energy Content

A gasoline-fueled internal combustion engine obtains energy from the combustion of gasoline's various hydrocarbons with oxygen from the ambient air, yielding carbon dioxide and water as exhaust. The combustion of octane, a representative species, performs the chemical reaction:

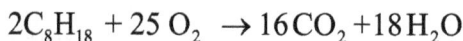

$$2C_8H_{18} + 25\,O_2 \rightarrow 16\,CO_2 + 18\,H_2O$$

Gasoline contains about 46.7 MJ/kg (127 MJ/US gal; 35.3 kWh/US gal; 13.0 kWh/kg; 120,405 BTU/US gal), quoting the lower heating value. Gasoline blends differ, and therefore actual energy content varies according to the season and producer by up to 1.75% more or less than the average. On average, about 74 L (19.5 US gal; 16.3 imp gal) of gasoline are available from a barrel of crude oil (about 46% by volume), varying with the quality of the crude and the grade of the gasoline. The remainder are products ranging from tar to naphtha.

A high-octane-rated fuel, such as liquefied petroleum gas (LPG), has an overall lower power output at the typical 10:1 compression ratio of an engine design optimized for gasoline fuel. An engine tuned for LPG fuel via higher compression ratios (typically 12:1) improves the power output. This is because higher-octane fuels allow for a higher compression ratio without knocking, resulting in a higher cylinder temperature, which improves efficiency. Also, increased mechanical efficiency is created by a higher compression ratio through the concomitant higher expansion ratio on the power stroke, which is by far the greater effect. The higher expansion ratio extracts more work from the high-pressure gas created by the combustion process. An Atkinson cycle engine uses the timing of the valve events to produce the benefits of a high expansion ratio without the disadvantages, chiefly detonation, of a high compression ratio. A high expansion ratio is also one of the two key reasons for the efficiency of diesel engines, along with the elimination of pumping losses due to throttling of the intake air flow.

The lower energy content of LPG by liquid volume in comparison to gasoline is due mainly to its lower density. This lower density is a property of the lower molecular weight of propane (LPG's chief component) compared to gasoline's blend of various hydrocarbon compounds with heavier molecular weights than propane. Conversely, LPG's energy content by weight is higher than gasoline's due to a higher hydrogen-to-carbon ratio.

Molecular weights of the representative octane combustion are C_8H_{18} 114, O_2 32, CO_2 44, H_2O 18; therefore 1 kg of fuel reacts with 3.51 kg of oxygen to produce 3.09 kg of carbon dioxide and 1.42 kg of water.

Octane Rating

Spark-ignition engines are designed to burn gasoline in a controlled process called deflagration. However, the unburned mixture may autoignite by pressure and heat alone, rather than igniting from the spark plug at exactly the right time, causing a rapid pressure rise which can damage the engine. This is often referred to as engine knocking or end-gas knock. Knocking can be reduced by increasing the gasoline's resistance to autoignition, which is expressed by its octane rating.

Octane rating is measured relative to a mixture of 2,2,4-trimethylpentane (an isomer of octane) and n-heptane. There are different conventions for expressing octane ratings, so the same physical fuel may have several different octane ratings based on the measure used. One of the best known is the research octane number (RON).

The octane rating of typical commercially available gasoline varies by country. In Finland, Sweden and Norway, 95 RON is the standard for regular unleaded gasoline and 98 RON is also available as a more expensive option.

In the United Kingdom, ordinary regular unleaded gasoline is sold at 95 RON (commonly available), premium unleaded gasoline is always 97 RON, and super-unleaded is usually 97–98 RON. However, both Shell and BP produce fuel at 102 RON for cars with high-performance engines, and in 2006 the supermarket chain Tesco began to sell super-unleaded gasoline rated at 99 RON.

In the United States, octane ratings in unleaded fuels vary between 85 and 87 AKI (91–92 RON) for regular, 89–90 AKI (94–95 RON) for mid-grade (equivalent to European regular), up to 90–94 AKI (95–99 RON) for premium (European premium).

	91	92	93	94	95	96	97	98	99	100	101	102
Scandinavia					regular			premium				
UK					regular		premium	super				high-performance
USA		regular		mid-grade			premium					

As South Africa's largest city, Johannesburg, is located on the Highveld at 1,753 metres (5,751 ft) above sea level, the Automobile Association of South Africa recommends 95-octane gasoline at low altitude and 93-octane for use in Johannesburg because "The higher the altitude the lower the air pressure, and the lower the need for a high octane fuel as there is no real performance gain".

Octane rating became important as the military sought higher output for aircraft engines in the late 1930s and the 1940s. A higher octane rating allows a higher compression ratio or supercharger boost, and thus higher temperatures and pressures, which translate to higher power output. Some scientists even predicted that a nation with a good supply of high-octane gasoline would have the advantage in air power. In 1943, the Rolls-Royce Merlin aero engine produced 1,320 horsepower (984 kW) using 100 RON fuel from a modest 27-liter displacement. By the time of Operation Overlord, both the RAF and USAAF were conducting some operations in Europe using 150 RON fuel (100/150 avgas), obtained by adding 2.5% aniline to 100-octane avgas. By this time the Rolls-Royce Merlin 66 was developing 2,000 hp using this fuel.

Additives

Antiknock Additives

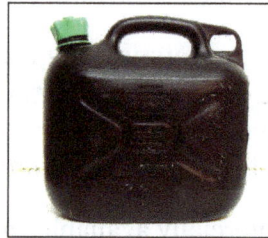

Plastic container for storing gasoline used in Germany.

Almost all countries in the world have phased out automotive leaded fuel. In 2011, six countries were still using leaded gasoline: Afghanistan, Myanmar, North Korea, Algeria, Iraq and Yemen. It was expected that by the end of 2013 those countries, too, would ban leaded gasoline, but this target was not met. Algeria replaced leaded with unleaded automotive fuel only in 2015. Different additives have replaced the lead compounds. The most popular additives include aromatic hydrocarbons, ethers and alcohol (usually ethanol or methanol). For technical reasons, the use of leaded additives is still permitted worldwide for the formulation of some grades of aviation gasoline such as 100LL, because the required octane rating would be technically infeasible to reach without the use of leaded additives.

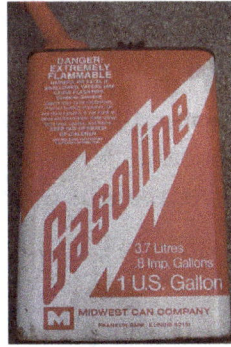

Gas can.

Tetraethyllead

Gasoline, when used in high-compression internal combustion engines, tends to autoignite or "detonate" causing damaging engine knocking (also called "pinging" or "pinking"). To address this problem, tetraethyllead (TEL) was widely adopted as an additive for gasoline in the 1920s. With the discovery of the seriousness of the extent of environmental and health damage caused by lead compounds, however, and the incompatibility of lead with catalytic converters, leaded gasoline was phased out in the United States beginning in 1973. By 1995, leaded fuel accounted for only 0.6 percent of total gasoline sales and under 2000 short tons (1814 t) of lead per year. From 1 January 1996, the U.S. Clean Air Act banned the sale of leaded fuel for use in on-road vehicles in the U.S. The use of TEL also necessitated other additives, such as dibromoethane.

European countries began replacing lead-containing additives by the end of the 1980s, and by the end of the 1990s, leaded gasoline was banned within the entire European Union. Reduction in the average lead content of human blood is believed to be a major cause for falling violent crime rates around the world, including in the United States and South Africa. A statistically significant correlation has been found between the usage rate of leaded gasoline and violent crime: taking into account a 22-year time lag, the violent crime curve virtually tracks the lead exposure curve.

Lead Replacement Petrol (Gasoline)

Lead replacement petrol (LRP) was developed for vehicles designed to run on leaded fuels and incompatible with unleaded fuels. Rather than tetraethyllead it contains other metals such as potassium compounds or methylcyclopentadienyl manganese tricarbonyl (MMT); these are purported to buffer soft exhaust valves and seats so that they do not suffer recession due to the use of unleaded fuel.

LRP was marketed during and after the phaseout of leaded motor fuels in the United Kingdom, Australia, South Africa and some other countries. Consumer confusion led to a widespread mistaken preference for LRP rather than unleaded, and LRP was phased out 8 to 10 years after the introduction of unleaded.

Leaded gasoline was withdrawn from sale in Britain after 31 December 1999, seven years after EEC regulations signaled the end of production for cars using leaded gasoline in member states. At this stage, a large percentage of cars from the 1980s and early 1990s which ran on leaded gasoline were still in use, along with cars which could run on unleaded fuel. However, the declining number of such cars on British roads saw many gasoline stations withdrawing LRP from sale by 2003.

MMT

Methylcyclopentadienyl manganese tricarbonyl (MMT) is used in Canada and the USA to boost octane rating. It also helps old cars designed for leaded fuel run on unleaded fuel without the need for additives to prevent valve problems. Its use in the United States has been restricted by regulations. Its use in the European Union is restricted by Article 8a of the Fuel Quality Directive following its testing under the Protocol for the evaluation of effects of metallic fuel-additives on the emissions performance of vehicles.

Fuel Stabilizers (Antioxidants and Metal Deactivators)

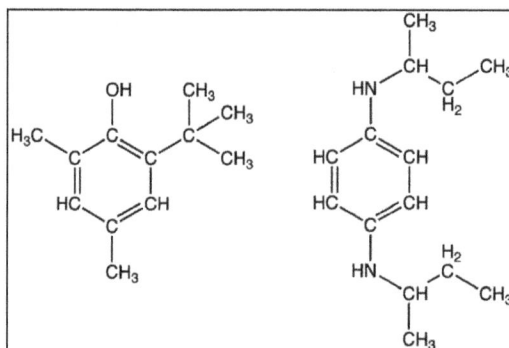

Substituted phenols and derivatives of phenylenediamine are common antioxidants used to inhibit gum formation in gasoline.

Gummy, sticky resin deposits result from oxidative degradation of gasoline during long-term storage. These harmful deposits arise from the oxidation of alkenes and other minor components in gasoline. Improvements in refinery techniques have generally reduced the susceptibility of gasolines to these problems. Previously, catalytically or thermally cracked gasolines were most susceptible to oxidation. The formation of gums is accelerated by copper salts, which can be neutralized by additives called metal deactivators.

This degradation can be prevented through the addition of 5–100 ppm of antioxidants, such as phenylenediamines and other amines. Hydrocarbons with a bromine number of 10 or above can be protected with the combination of unhindered or partially hindered phenols and oil-soluble strong amine bases, such as hindered phenols. "Stale" gasoline can be detected by a colorimetric enzymatic test for organic peroxides produced by oxidation of the gasoline.

Gasolines are also treated with metal deactivators, which are compounds that sequester (deactivate) metal salts that otherwise accelerate the formation of gummy residues. The metal impurities might arise from the engine itself or as contaminants in the fuel.

Detergents

Gasoline, as delivered at the pump, also contains additives to reduce internal engine carbon buildups, improve combustion and allow easier starting in cold climates. High levels of detergent can be found in Top Tier Detergent Gasolines. The specification for Top Tier Detergent Gasolines was developed by four automakers: GM, Honda, Toyota, and BMW. According to the bulletin, the minimal U.S. EPA requirement is not sufficient to keep engines clean. Typical detergents include alkylamines and alkyl phosphates at the level of 50–100 ppm.

Dyes

Though gasoline is a naturally colorless liquid, many gasolines are dyed in various colors to indicate their composition and acceptable uses. In Australia, the lowest grade of gasoline (RON 91) was dyed a light shade of red/orange and is now the same colour as the medium grade (RON 95) and high octane (RON 98) which are dyed yellow. In the United States, aviation gasoline (avgas) is dyed to identify its octane rating and to distinguish it from kerosene-based jet fuel, which is clear. In Canada, the gasoline for marine and farm use is dyed red and is not subject to sales tax.

Oxygenate Blending

Oxygenate blending adds oxygen-bearing compounds such as MTBE, ETBE, TAME, TAEE, ethanol and biobutanol. The presence of these oxygenates reduces the amount of carbon monoxide and unburned fuel in the exhaust. In many areas throughout the U.S., oxygenate blending is mandated by EPA regulations to reduce smog and other airborne pollutants. For example, in Southern California, fuel must contain 2% oxygen by weight, resulting in a mixture of 5.6% ethanol in gasoline. The resulting fuel is often known as reformulated gasoline (RFG) or oxygenated gasoline, or in the case of California, California reformulated gasoline. The federal requirement that RFG contain oxygen was dropped on 6 May 2006 because the industry had developed VOC-controlled RFG that did not need additional oxygen.

MTBE was phased out in the U.S. due to groundwater contamination and the resulting regulations and lawsuits. Ethanol and, to a lesser extent, the ethanol-derived ETBE are common substitutes. A common ethanol-gasoline mix of 10% ethanol mixed with gasoline is called gasohol or E10, and an ethanol-gasoline mix of 85% ethanol mixed with gasoline is called E85. The most extensive use of ethanol takes place in Brazil, where the ethanol is derived from sugarcane. In 2004, over 3.4 billion US gallons (2.8 billion imp gal; 13 million m³) of ethanol was produced in the United States for fuel use, mostly from corn, and E85 is slowly becoming available in much of the United States, though many of the relatively few stations vending E85 are not open to the general public.

The use of bioethanol and bio-methanol, either directly or indirectly by conversion of ethanol to bio-ETBE, or methanol to bio-MTBE is encouraged by the European Union Directive on the Promotion of the use of biofuels and other renewable fuels for transport. Since producing bioethanol from fermented sugars and starches involves distillation, though, ordinary people in much of Europe cannot legally ferment and distill their own bioethanol at present (unlike in the U.S., where getting a BATF distillation permit has been easy since the 1973 oil crisis).

Safety

HAZMAT class 3 gasoline.

Environmental Considerations

Combustion of 1 U.S. gallon (3.8 L) of gasoline produces 8.74 kilograms (19.3 lb) of carbon dioxide (2.3 kg/l), a greenhouse gas.

The main concern with gasoline on the environment, aside from the complications of its extraction and refining, is the effect on the climate through the production of carbon dioxide. Unburnt gasoline and evaporation from the tank, when in the atmosphere, reacts in sunlight to produce photochemical smog. Vapor pressure initially rises with some addition of ethanol to gasoline, but the increase is greatest at 10% by volume. At higher concentrations of ethanol above 10%, the vapor pressure of the blend starts to decrease. At a 10% ethanol by volume, the rise in vapor pressure may potentially increase the problem of photochemical smog. This rise in vapor pressure could be mitigated by increasing or decreasing the percentage of ethanol in the gasoline mixture.

The chief risks of such leaks come not from vehicles, but from gasoline delivery truck accidents and leaks from storage tanks. Because of this risk, most (underground) storage tanks now have extensive measures in place to detect and prevent any such leaks, such as monitoring systems (Veeder-Root, Franklin Fueling).

Production of gasoline consumes 0.63 gallons of water per mile driven.

Toxicity

The safety data sheet for a 2003 Texan unleaded gasoline shows at least 15 hazardous chemicals occurring in various amounts, including benzene (up to 5% by volume), toluene (up to 35% by volume), naphthalene (up to 1% by volume), trimethylbenzene (up to 7% by volume), methyl *tert*-butyl ether (MTBE) (up to 18% by volume, in some states) and about ten others. Hydrocarbons in gasoline generally exhibit low acute toxicities, with LD50 of 700–2700 mg/kg for simple aromatic compounds. Benzene and many antiknocking additives are carcinogenic.

People can be exposed to gasoline in the workplace by swallowing it, breathing in vapors, skin contact, and eye contact. Gasoline is toxic. The National Institute for Occupational Safety and Health (NIOSH) has also designated gasoline as a carcinogen. Physical contact, ingestion or inhalation can cause health problems. Since ingesting large amounts of gasoline can cause permanent damage to major organs, a call to a local poison control center or emergency room visit is indicated.

Contrary to common misconception, swallowing gasoline does not generally require special emergency treatment, and inducing vomiting does not help, and can make it worse. According to poison specialist Brad Dahl, "even two mouthfuls wouldn't be that dangerous as long as it goes down to your stomach and stays there or keeps going." The US CDC's Agency for Toxic Substances and Disease Registry says not to induce vomiting, lavage, or administer activated charcoal.

Inhalation for Intoxication

Inhaled (huffed) gasoline vapor is a common intoxicant. Users concentrate and inhale gasoline vapour in a manner not intended by the manufacturer to produce euphoria and intoxication. Gasoline inhalation has become epidemic in some poorer communities and indigenous groups in Australia, Canada, New Zealand, and some Pacific Islands. The practice is thought to cause severe organ damage, including mental retardation.

In Canada, Native children in the isolated Northern Labrador community of Davis Inlet were the focus of national concern in 1993, when many were found to be sniffing gasoline. The Canadian and provincial Newfoundland and Labrador governments intervened on a number of occasions, sending many children away for treatment. Despite being moved to the new community of Natuashish in 2002, serious inhalant abuse problems have continued. Similar problems were reported in Sheshatshiu in 2000 and also in Pikangikum First Nation. In 2012, the issue once again made the news media in Canada.

Australia has long faced a petrol (gasoline) sniffing problem in isolated and impoverished aboriginal communities. Although some sources argue that sniffing was introduced by United States servicemen stationed in the nation's Top End during World War II or through experimentation by 1940s-era Cobourg Peninsula sawmill workers, other sources claim that inhalant abuse (such as glue inhalation) emerged in Australia in the late 1960s. Chronic, heavy petrol sniffing appears to occur among remote, impoverished indigenous communities, where the ready accessibility of petrol has helped to make it a common substance for abuse.

In Australia, petrol sniffing now occurs widely throughout remote Aboriginal communities in the Northern Territory, Western Australia, northern parts of South Australia and Queensland. The number of people sniffing petrol goes up and down over time as young people experiment or sniff occasionally. "Boss", or chronic, sniffers may move in and out of communities; they are often responsible for encouraging young people to take it up. In 2005, the Government of Australia and BP Australia began the usage of Opal fuel in remote areas prone to petrol sniffing. Opal is a non-sniffable fuel (which is much less likely to cause a high) and has made a difference in some indigenous communities.

Flammability

Like other hydrocarbons, gasoline burns in a limited range of its vapor phase and, coupled with its volatility, this makes leaks highly dangerous when sources of ignition are present. Gasoline has a lower explosive limit of 1.4% by volume and an upper explosive limit of 7.6%. If the concentration is below 1.4%, the air-gasoline mixture is too lean and does not ignite. If the concentration is above 7.6%, the mixture is too rich and also does not ignite. However, gasoline vapor rapidly mixes and spreads with air, making unconstrained gasoline quickly flammable.

Uncontrolled burning of gasoline produces large quantities of soot and carbon monoxide.

Carbon Dioxide Production

About 19.64 pounds (8.91 kg) of carbon dioxide (CO_2) are produced from burning 1 U.S. gallon (3.8 liters) of gasoline that does not contain ethanol (2.36 kg/L). About 22.38 pounds (10.15 kg) of CO_2 are produced from burning one US gallon of diesel fuel (2.69 kg/l).

The U.S. EIA estimates that U.S. motor gasoline and diesel (distillate) fuel consumption for transportation in 2015 resulted in the emission of about 1,105 million metric tons of CO_2 and 440 million metric tons of CO_2, respectively, for a total of 1,545 million metric tons of CO_2. This total was equivalent to 83% of total U.S. transportation-sector CO_2 emissions and equivalent to 29% of total U.S. energy-related CO_2 emissions in 2015.

Most of the retail gasoline now sold in the United States contains about 10% fuel ethanol (or E10) by volume. Burning a gallon of E10 produces about 17.68 pounds (8.02 kg) of CO_2 that is emitted from the fossil fuel content. If the CO_2 emissions from ethanol combustion are considered, then about 18.95 pounds (8.60 kg) of CO_2 are produced when a gallon of E10 is combusted. About 12.73 pounds (5.77 kg) of CO_2 are produced when a gallon of pure ethanol is combusted.

Kerosene

Kerosene, also called paraffin or paraffin oil is a flammable hydrocarbon liquid commonly used as a fuel. Kerosene is typically pale yellow or colourless and has a not-un-pleasant characteristic odour. It is obtained from petroleum and is used for burning in kerosene lamps and domestic heaters or furnaces, as a fuel or fuel component for jet engines, and as a solvent for greases and insecticides.

Discovered by Canadian physician Abraham Gesner in the late 1840s, kerosene was initially manufactured from coal tar and shale oils. However, following the drilling of the first oil well in Pennsylvania by E.L. Drake in 1859, petroleum quickly became the major source of kerosene. Because of its use in lamps, kerosene was the major refinery product for several decades until the advent of the electric lamp reduced its value for lighting. Production further declined as the rise of the automobile established gasoline as an important petroleum product. Nevertheless, in many parts of the world, kerosene is still a common heating and cooking fuel as well as a fuel for lamps. Standard commercial jet fuel is essentially a high-quality straight-run kerosene, and many military jet fuels are blends based on kerosene.

Chemically, kerosene is a mixture of hydrocarbons. The chemical composition depends on its source, but it usually consists of about 10 different hydrocarbons, each containing 10 to 16 carbon atoms per molecule. The main constituents are saturated straight-chain and branched-chain paraffins, as well as ring-shaped cycloparaffins (also known as naphthenes). Kerosene is less volatile than gasoline. Its flash point (the temperature at which it will generate a flammable vapour near its surface) is 38 °C (100 °F) or higher, whereas that of gasoline is as low as −40 °C (−40 °F). This property makes kerosene a relatively safe fuel to store and handle.

With a boiling point between about 150 and 300 °C (300–575 °F), kerosene is considered to be one of the so-called middle distillates of crude oil, along with diesel fuel. It can be produced as "straight-run kerosene," separated physically from the other crude oil fractions by distillation, or it can be produced as "cracked kerosene," by chemically decomposing, or cracking, heavier portions of the oil at elevated temperatures.

References

- Sami Matar and Lewis F. Hatch (2001). Chemistry of Petrochemical Processes. Gulf Professional Publishing. ISBN 0-88415-315-0

- Paraffin-wax, science: britannica.com, Retrieved 3 July, 2019

- Alvi, Moin ud-Din. "Aerosol Propellant | Aerosol Propellant Gas | Aerosol Supplies Dubai – Brothers Gas". Www.brothersgas.com. Archived from the original on 30 December 2016. Retrieved 14 June 2016

- Werner Dabelstein, Arno Reglitzky, Andrea Schütze and Klaus Reders "Automotive Fuels" in Ullmann's Encyclopedia of Industrial Chemistry 2007, Wiley-VCH, Weinheim. Doi:10.1002/14356007. a16_719.pub2

- Crude-oil, science: ritannica.com, Retrieved 4 August, 2019

- "SNOX Process: A Success Story" Archived 2009-07-21 at the Wayback Machine, energystorm. us. Cited therein: "Schoolbook, Chemistry 2000, Helge Mygind, ISBN 87-559-0992-2".

- Diesel-fuel, technology: britannica.com, Retrieved 5 January, 2019

- "Petrolatum (white)". Inchem.org. International Programme on Chemical Safety and the Commission of the European Communities. March 2002. Retrieved August 5,2011

- Kerosene, science: britannica.com, Retrieved 6 February 2019

6

Environmental Impacts of Petroleum Production

Due to the toxicity of petroleum it causes various negative environmental impacts such as air pollution, acid rain and illnesses in humans. The topics elaborated in this chapter will help in gaining a better perspective about the environmental impacts of different processes related to petroleum production such as crude oil and natural gas production, hydraulic fracturing, and the oil shale industry.

Environmental and Economic Impacts of Crude Oil and Natural Gas Production

Oil and gas have remained the lifeblood of the world economy for over one hundred years accounting for over half of mankind's primary energy supply. These high energy density and easily available fossil fuels have played important roles in some of the biggest industries like chemicals, transport, power, petrochemicals etc. The availability of cheap, abundant energy lifts nations out of poverty and at such, energy security has become national priority for most nations. Crude oil and natural gas supply has become very important especially in the face of rising demand for energy for comfort and technological development.

The total measure of economic effects of crude oil and natural gas production on the host nation or community especially for developing countries could be best described by the impacts: direct, indirect and induced. The direct impacts are measured as the jobs, labour income and value added to the oil and gas industry whereas indirect impacts are measured with the same yardstick but occurring across the supply chain due to crude oil and natural gas production activities. Induced impacts are measured as jobs labour income, and value addition resulting from household spending of labour and proprietor's income earned either directly or indirectly from crude oil and natural gas production activities.

Crude oil and natural gas production activities have been found to make enormous economic contributions that benefit both the host nations and the citizenry. Some of the ways through which crude oil and natural gas production contribute to the economy include.

Taxes

Oil and gas companies, involved in crude oil and natural gas production, pay billions of dollars in taxes to the government of their host countries every year. These funds help pay for important government services, such as education, health care and provide infrastructure that benefit the citizens of the country. In 2013, Canada's oil and natural gas industry paid a total of $18 billion to federal, provincial and local governments in the form of taxes and royalties.

Oil and Gas Royalties

Royalty is the share the government receives from companies producing crude oil and natural gas from the nation's reserves. The amount is usually dependent on the volume of crude oil or natural gas produced.

Employment and Job Creation

The oil and gas industry employs millions of people all over the world. These are usually high-paying jobs that make a great percentage of them live above the average income, spend within the community and pay taxes to the governments.

Gross Regional Product (GRP)

The total change in value added generated by direct spending. GRP is conceptually the same type of measure as gross domestic product (GDP), which is also a measure of value added and indicates the market value of goods and services, at purchaser prices, produced by all economic resources located in the country. Crude oil and natural gas production have been found to increase the GRP of the states and in turn the GDP of the country.

Local Expenditure on Goods and Services

The periodic injection of purchasing power through its local expenditure on goods and services by the oil and gas production industry is another way of contributing to the economy of the host country. Payments to local contractors for goods and services and for direct purchases has the capacity to stimulate the economy and also exert secondary influences, through multiplier process on the level of output and employment in other related sections of the economy.

Provision of Foreign Exchange Reserves

Since crude oil and natural gas are sold in international markets, producing countries has the potential to earn and save foreign exchange in reserves. This puts the country's economy in healthy position and gives her the capacity to finance the foreign exchange cost of any development program.

Contribution to Power Supply and Public Utilities

Natural gas could be used to power turbines for the generation of electricity. Associated gas is still being flared today in many oil fields in developing countries: this could be turned to power and used for industrialization of the producing community. In some cases, producing companies have supplied electricity to the host communities to aid development.

Investment

Most oil and gas companies are quoted at the various stock exchange markets. This makes it possible for citizens to invest in these stocks and enjoy the privilege of the innovation, growth and dividends associated with such investment.

Crude oil and natural gas production is done either on land or on water. The area occupied for this purpose would have been used for the purpose of farming and fishing respectively. In most communities where oil and gas production activities are being carried out, the traditional occupations of the people prior to oil and gas discovery are being abandoned. The country, in some cases, may even resort to total reliance on the production of oil and natural gas for economic growth leading to a mono product economy. For example, Nigeria is currently in this dilemma as over 70% of the country's earning is crude oil and natural gas dependent.

Although a great deal of the expertise required to produce crude oil and natural gas is sourced from locations far from the field; in some cases overseas, the industry most of the times provides the host and surrounding communities with employment for unskilled and low skilled labour. There are opportunities to learn from the best hands in the industry and get exposed to the most recent technologies on the various operations. In cases where the production of crude oil and natural gas is done in onshore locations, minor sub-contractors from the host and adjoining communities are usually given priority after pre-qualification. In all, the communities benefit from employment, provision of infrastructure (in the form of corporate social responsibility from the companies), award of scholarships, shopping from employees of the company etc.

Environmental Impacts of Crude Oil and Natural Gas Production

Environmental impacts that occur during production of crude oil and natural gas would mostly occur from long-term habitat change within the oil and gas field, production activities (including facility component maintenance or replacement), waste management (e.g produced water), noise (e.g from well operations, compressor or pump stations, flare stack, vehicle and equipment), the presence of workers and potential spills. These activities could potentially impact on the resources.

Noise

The main sources of noise during the production of crude oil and natural gas would include compressor and pumping stations, producing wells (including occasional flaring), and vehicle traffic. Compressor stations produce noise levels between 64 and 86 dBA at the station to between 58 and 75 dBA at about 1 mile (1.6 kilometers) from the station. The primary impacts from noise would be localized disturbance to wildlife, recreationists, and residents. Noise associated with cavitation is a major concern for landowners, livestock, and wildlife.

Air Quality

The primary emission sources during the production of crude oil and natural gas would include compressor and pumping station operations, vehicle traffic, production well operations, separation of oil and gas phases, and on-site storage of crude oil. Emissions would include volatile organic compound (VOCs), nitrogen oxides, sulfur dioxide, carbon monoxide, benzene, toluene, ethylbenzene, xylenes, polycyclic aromatic hydrocarbons (PAHs), hydrogen sulfide, particulates, ozone, and methane. Venting or flaring of natural gas (methane) may occur during oil production, well testing, oil and gas processing, cavitation, well leaks, and pipeline maintenance operations. Methane is a major greenhouse gas. Air pollution during oil and gas production may cause health effects and reduce visibility.

Cultural Resources

Production of crude oil and natural gas could also impact on the cultural resources by unauthorized collection of artifacts and the alteration of visual image. The presence of the aboveground structures alters the associated landscape component of the cultural resources. Damage to localities caused through off-highway vehicle (OHV) and the potential for indirect impacts (e.g, vandalism and unauthorized collecting) also exist.

Ecological Resources

The adverse impacts to ecological resources during production of crude oil and natural gas could occur from: disturbance of wildlife from noise and human activity; exposure of biota to contaminants; and mortality of biota from colliding with aboveground facilities or vehicles. The presence of production wells, ancillary facilities and access road reduces the habitat quality, disturbs the biota and thus affects ecological resources. The presence of an oil or gas field could also interfere with migratory and other behaviors of some wildlife. Discharge of produced water inappropriately onto soil or into surface water bodies can result in salinity levels too high to sustain plant growth. Wildlife is always prone to contact with petroleum-based products and other contaminants in reserve pits and water management facilities. They can become entrapped in the oil and drown, ingest toxic quantities of oil by preening (birds) or licking their fur (mammals); or succumb to cold stress if the oil damages the insulation provided by feathers or fur.

In locations where naturally occurring radioactive material (NORM)-bearing produced water and solid wastes are generated, mismanagement of these wastes can result in radiological contamination of soils or surface water bodies.

Hazardous Materials and Waste Management

Industrial wastes are generated during routine operations (lubricating oils, hydraulic fluids, coolants, solvents, and cleaning agents). These wastes are typically placed in containers, characterized, labeled and possibly stored briefly before being transported by a licensed hauler to an appropriate permitted off-site disposal facility as a standard practice. Impacts could result if these wastes were not properly handled and were released to the environment. Environmental contamination could occur from accidental spills of herbicides or, more significantly, oil. Chemicals in open pits used to store wastes may pose a threat to wildlife and livestock. "Fracking" fluids can contain potentially toxic substances such as diesel fuel (which contains benzene, ethylbenzene, toluene, xylenes, naphthalene, and other chemicals), PAHs, methanol, formaldehyde, ethylene glycol, glycol ethers, hydrochloric acid, and sodium hydroxide. Sand separated from produced water must be properly disposed as it is often contaminated with oil, trace amounts of metals, or other naturally occurring constituents. Production could also cause accumulation of large volumes of scale and sludge wastes inside pipelines and storage vessels. These wastes may be transported to offsite disposal facilities. Produced water can become a significant waste stream during the production of crude oil and natural gas. Regulations govern the disposal of this waste stream; the majority of it is disposed by underground injection either in disposal wells or, in mature producing fields, in enhanced oil recovery wells (i.e, wells through which produced water and other materials are injected into a producing formation in order to increase formation pressure and production). In some locations, produced water may carry NORM to the surface.

Health and Safety

Possible impacts to public health and safety during production include accidental injury or death to workers and, to a lesser extent, the public (e.g, from an OHV collisions with project components or vehicle collisions with oil or gas workers). Health impacts could result from water contamination, dust and other air emissions, noise, soil contamination, and stress (e.g, associated with living near an industrial zone). Potential fires and explosions would cause safety hazards. Cavitation could ignite grass fires. Increased or reckless driving by oil or gas workers would also create safety hazards. In addition, health and safety issues include working in potential weather extremes and possible contact with natural hazards, such as uneven terrain and dangerous plants, animals, or insects.

Land use

Land use impacts during the production of crude oil and natural gas would be an extension of those that occurred during the drilling/development phase. Although it is possible for farmers or fisher men to carry out activities around the well locations, restrictions would always exist.

Paleontological Resources

The existence of access roads creates a threat to paleontological resources during oil and gas production allowing for unauthorized collection of fossils.

Socioeconomics

Although new jobs and businesses would be created and royalties and taxes paid to land owners and government, there is a potential negative impact on the value of properties located in the proximity of oil and gas field. This effect increases as the number of wells increase.

Soils and Geological Resources

The main impact from production would be the depletion of recoverable oil and gas reserves. Possible geological hazards (earthquakes, landslides, and subsidence) could be activated by oil and gas extraction activities. Although it is rare, the injection of produced water in disposal wells could trigger localized seismic activity.

Transportation

The impact of crude oil and natural gas production to transportation would be basically due to the daily vehicular movement of light trucks and cars used for surveillance and movement of materials. Heavy truck traffic would be limited to periodic visits to a well site for workovers and formation treatment.

Water Resources

During the life of a production well, the integrity of the well casing and cement will determine the potential for adverse impacts to groundwater. If subsurface formations are not sealed off by the well casing and cement, aquifers can be impacted by other non-potable formation waters. Hydraulic fracturing fluids have the potential to contaminate groundwater drinking reservoirs. Stimulation fluids may penetrate away from the fracture and into surrounding formation. When stimulation ceases and production resumes, these chemicals may not be completely recovered and pumped back into the wellbore, and, if mobile, may be available to migrate through an aquifer. Most produced water is unfit for domestic or agricultural purposes (e.g, it is extremely salty or contains NORM or toxic compounds). If it is disposed of by release to the surface without treatment, it can cause soil and surface water contamination. The majority of produced water is disposed via injection in disposal wells or enhanced recovery wells.

Toxic Compounds Associated with Crude Oil and Natural Gas Production

Although crude oil and natural gas have played great roles in transportation, electricity production, industrial power, military and medical applications; they are also raw materials for production of weapons of war, reasons behind some political unrest and human rights violations, and have caused environmental degradation and some human diseases.

The operations leading to the production of crude oil and natural gas could lead to the emission of some compounds that constitute risk to the environment and public health. Some specific compounds associated with oil and gas production, their sources and potential effect on health are:

Benzene

This is a well-established carcinogen with specific links to leukemia as well as breast and urinary tract cancers. Exposure to benzene reduces red and white blood cells production in the bone marrow, decreases auto-immune cell function (T-cell and B-cells), and has been linked to sperm-head abnormalities and generalised chromosome aberrations. Benzene is a commonly used petrochemical solvent in the production of crude oil and natural gas especially in fracking operations.

Sulphur IV Oxide

This compound combines with the oxides of nitrogen to form particulate matter which has been known to contribute to serious health problems including cancer of the lung and cardiopulmonary mortality. Exposure of children to SO_2 even at lower levels over time has been found to cause asthma. Sulphur IV Oxide is released by combustion of crude oil and natural gas at the flare stack during production.

Oxides of Nitrogen (NO_x)

NO_x are involved in the formation of particulate matter and also contribute directly to thousands of hospitalisations, heart attacks and deaths annually. Inhalation of NO_x is associated with emphysema and bronchitis. Flaring is the chief source of NO_x during the production of crude oil and natural gas.

Formaldehyde

This compound is another carcinogen with known links to leukemia and rare nasopharyngeall cancers. Studies have linked spontaneous abortions, congenital malformations, low birth weights, infertility and endometriosis to formaldehyde exposure. It also contributes to ground-level ozone and is commonly used in fracking.

Polycyclic Aromatic Hydrocarbons (PAH)

This is an entire class of toxic chemicals, linked together by their unique chemical structure and reactive properties. Many PAHs are known human carcinogens and genetic mutagens. In addition, there are particular prenatal health risks: prenatal exposure to PAHs is linked to childhood asthma, low birth weight, adverse birth outcomes including heart malformations and DNA damage. Recent studies link exposure to early childhood depression. The main source of PAHs during production is spills.

Silica

Crystalline silica is a known human carcinogen; breathing silica dust can lead to silicosis, a form of lung disease with no cure. Silica is commonly used, in huge amounts, during fracking operations. Each stage of the process requires hundreds of thousands of pounds of silica quartz–containing sand. Millions of pounds may be used for a single well.

Radon

This is a colourless, odourless, tasteless radioactive gas which causes lung cancer. It is the second largest cause of lung cancer in the U.S. after cigarette smoking. There is no known threshold below which radon exposures carries no risk. Radon is released into groundwater and air during natural gas production especially by fracking.

Hydrofluoric Acid (HF)/Hydrogen Fluoride

Hydrofluoric acid (HF) is one of the most dangerous acids known. HF can immediately damage lungs, leading to chronic lung disease; contact on skin penetrates to deep tissue, including bone, where it alters cellular structure. HF can be fatal if inhaled, swallowed, or absorbed through skin. Hydrofluoric acid is used for well simulation during crude oil and natural gas production.

Hydrogen Sulphide (H_2S)

This is a colourless gas with a rotten egg smell but odourless at concentrations above 150ppb as it quickly impairs olfactory sense. It is heavier than air, very poisonous, corrosive, flammable and explosive. As the concentration of H_2S being inhaled increases, it could lead to pulmonary edema, respiratory paralysis, collapse and even death. H_2S could be released from leaking Christmas trees at the wellhead, pumps, piping, separation device, oil storage tanks, waste water storage tanks, intentional venting and flaring during production.

Carbon Dioxide, Carbon Monoxide, Volatile Organic Compounds and Particulate Matter

These compounds are generated by burning of crude oil and natural gas during production process have negative impacts on the environment and human health:

Carbon dioxide is a greenhouse gas and a source of global warming; VOCs contribute to ground-level ozone, which irritates and damages the lungs; PM results in hazy conditions in cities and scenic areas, and, along with ozone, contributes to asthma and chronic bronchitis, especially in children and the elderly.

Methane

Methane is the major constituent of natural gas. It is a colourless gas which is odourless at low concentration but has a sweet smell high concentrations. It is a greenhouse gas that has very high global warming potential (about 21 times that of carbon dioxide). It could lead to suffocation and death at high concentration in enclosed space. It is released into the atmosphere by venting operations and leaking valves during crude oil and natural gas production.

Strategies for the Protection of the Environment

The release of these toxic compounds into the environment is principally caused by flaring, venting, improper cementing or sealing of well bore, lack of maintenance of production facilities, inefficient produced water and solid waste management scheme, poor handling of crude oil and natural gas leading to spills and leaks and non-adherence to regulations. In order to limit the release of these toxic compounds into the environment, the following strategies have been proposed:

- Commercialization of associated gas, application of new technologies, re-injection of associated gas, regulations, legislations and promotion of best practices are some of the ways used to reduce flaring and venting in crude oil and natural gas production processes.

- Proper management of well drilling and workover processes as well as integrity checks on oil and gas wells before abandonment will help check seepage of oil and gas into ground water while improved well control procedure will prevent blowouts.

- Valves and pumps are common in oil and gas production processes. Operation and maintenance plans are usually prepared for the valves, pumps and other equipment to ensure they are in good condition; and not leaking out fluids.

- Produced water management facilities are usually incorporated into most production process. It could be either produced water injection system or treatment to an allowable limit before discharge into water body, in the case of offshore locations.

- Good and efficient crude oil and natural gas handling procedure is developed for the production facility making sure that all regulations with regards to production of crude oil and natural gas are complied with.

Environmental Impact of Hydraulic Fracturing

The environmental impact of hydraulic fracturing is related to land use and water consumption, air emissions, including methane emissions, brine and fracturing fluid leakage, water contamination, noise pollution, and health. Water and air pollution are the biggest risks to human health from hydraulic fracturing. Research is underway to determine if human health has been affected, and adherence to regulation and safety procedures is required to avoid negative impacts.

Hydraulic fracturing fluids include proppants and other substances, which may include toxic chemicals. In the United States, such additives may be treated as trade secrets by companies who use them. Lack of knowledge about specific chemicals has complicated efforts to develop risk management policies and to study health effects. In other jurisdictions, such as the United Kingdom, these chemicals must be made public and their applications are required to be nonhazardous.

Water usage by hydraulic fracturing can be a problem in areas that experience water shortage. Surface water may be contaminated through spillage and improperly built and maintained waste pits, in jurisdictions where these are permitted. Further, ground water can be contaminated if fracturing fluids and formation fluids are able to escape during hydraulic fracturing. However, the possibility of groundwater contamination from the fracturing fluid upward migration is negligible, even in a long-term period. Produced water, the water that returns to the surface after hydraulic fracturing, is managed by underground injection, municipal and commercial wastewater treatment, and reuse in future wells. There is potential for methane to leak into ground water and the air, though escape of methane is a bigger problem in older wells than in those built under more recent legislation.

Hydraulic fracturing causes induced seismicity called microseismic events or micro-earthquakes. The magnitude of these events is too small to be detected at the surface, being of magnitude M-3 to M-1 usually. However, fluid disposal wells (which are often used in the USA to dispose of polluted waste from several industries) have been responsible for earthquakes up to 5.6M in Oklahoma and other states.

Governments worldwide are developing regulatory frameworks to assess and manage environmental and associated health risks, working under pressure from industry on the one hand, and from anti-fracking groups on the other. In some countries like France a precautionary approach has been favored and hydraulic fracturing has been banned. The United Kingdom's regulatory framework is based on the conclusion that the risks associated with hydraulic fracturing are manageable if carried out under effective regulation and if operational best practices are implemented.

Air Emissions

A report for the European Union on the potential risks was produced in 2012. Potential risks are "methane emissions from the wells, diesel fumes and other hazardous pollutants, ozone precursors or odours from hydraulic fracturing equipment, such as compressors, pumps, and valves". Also gases and hydraulic fracturing fluids dissolved in flowback water pose air emissions risks. One study measured various air pollutants weekly for a year surrounding the development of a newly fractured gas well and detected nonmethane hydrocarbons, methylene chloride (a toxic solvent), and polycyclic aromatic hydrocarbons. These pollutants have been shown to affect fetal outcomes.

The relationship between hydraulic fracturing and air quality can influence acute and chronic respiratory illnesses, including exacerbation of asthma (induced by airborne particulates, ozone and exhaust from equipment used for drilling and transport) and COPD. For example, communities overlying the Marcellus shale have higher frequencies of asthma. Children, active young adults who spend time outdoors, and the elderly are particularly vulnerable. OSHA has also raised concerns about the long-term respiratory effects of occupational exposure to airborne silica at hydraulic fracturing sites. Silicosis can be associated with systemic autoimmune processes.

"In the UK, all oil and gas operators must minimise the release of gases as a condition of their licence from the Department of Energy and Climate Change (DECC). Natural gas may only be vented for safety reasons."

Also transportation of necessary water volume for hydraulic fracturing, if done by trucks, can cause emissions. Piped water supplies can reduce the number of truck movements necessary.

A report from the Pennsylvania Dept of Environmental Protection indicated that there is little potential for radiation exposure from oil and gas operations.

Air pollution is of particular concern to workers at hydraulic fracturing well sites as the chemical emissions from storage tanks and open flowback pits combine with the geographically compounded air concentrations from surrounding wells. Thirty seven percent of the chemicals used in hydraulic fracturing operations are volatile and can become airborne.

Researchers Chen and Carter from the Department of Civil and Environmental Engineering, University of Tennessee, Knoxville used atmospheric dispersion models (AERMOD) to estimate the potential exposure concentration of emissions for calculated radial distances of 5 m to 180m from emission sources. The team examined emissions from 60,644 hydraulic fracturing wells and found results showed the percentage of wells and their potential acute non-cancer, chronic non-cancer, acute cancer, and chronic cancer risks for exposure to workers were 12.41%, 0.11%, 7.53%, and 5.80%, respectively. Acute and chronic cancer risks were dominated by emissions from the chemical storage tanks within a 20 m radius.

Climate Change

Whether natural gas produced by hydraulic fracturing causes higher well-to-burner emissions than gas produced from conventional wells is a matter of contention. Some studies have found that hydraulic fracturing has higher emissions due to methane re-leased during completing wells as some gas returns to the surface, together with the fracturing fluids. Depending on their treatment, the well-to-burner emissions are 3.5%–12% higher than for conventional gas.

A debate has arisen particularly around a study by professor Robert W. Howarth find-ing shale gas significantly worse for global warming than oil or coal. Other research-ers have criticized Howarth's analysis, including Cathles *et al.,* whose estimates were substantially lower." A 2012 industry funded report co-authored by researchers at the United States Department of Energy's National Renewable Energy Laboratory found emissions from shale gas, when burned for electricity, were "very similar" to those from so-called "conventional well" natural gas, and less than half the emissions of coal.

Several studies which have estimated lifecycle methane leakage from natural gas de-velopment and production have found a wide range of leakage rates. According to the Environmental Protection Agency's Greenhouse Gas Inventory, the methane leakage rate is about 1.4%. A 16-part assessment of methane leakage from natural gas produc-tion initiated by the Environmental Defense Fund found that fugitive emissions in key stages of the natural gas production process are significantly higher than estimates in the EPA's national emissions inventory, with a leakage rate of 2.3 percent of overall natural gas output.

Water Consumption

Massive hydraulic fracturing typical of shale wells uses between 1.2 and 3.5 million US gallons (4,500 and 13,200 m^3) of water per well, with large projects using up to 5 mil-lion US gallons (19,000 m^3). Additional water is used when wells are refractured. An average well requires 3 to 8 million US gallons (11,000 to 30,000 m^3) of water over its lifetime. According to the Oxford Institute for Energy Studies, greater volumes of frac-turing fluids are required in Europe, where the shale depths average 1.5 times greater than in the U.S. Whilst the published amounts may seem large, they are small in com-parison with the overall water usage in most areas. A study in Texas, which is a water shortage area, indicates "Water use for shale gas is <1% of statewide water withdrawals; however, local impacts vary with water availability and competing demands."

A report by the Royal Society and the Royal Academy of Engineering shows the usage expected for hydraulic fracturing a well is approximately the amount needed to run a 1,000 MW coal-fired power plant for 12 hours. A 2011 report from the Tyndall Centre es-timates that to support a 9 billion cubic metres per annum (320×10^9 cu ft/a) gas produc-tion industry, between 1.25 to 1.65 million cubic metres (44×10^6 to 58×10^6 cu ft) would be needed annually, which amounts to 0.01% of the total water abstraction nationally.

Concern has been raised over the increasing quantities of water for hydraulic fracturing in areas that experience water stress. Use of water for hydraulic fracturing can divert water from stream flow, water supplies for municipalities and industries such as power generation, as well as recreation and aquatic life. The large volumes of water required for most common hydraulic fracturing methods have raised concerns for arid regions, such as the Karoo in South Africa, and in drought-prone Texas, in North America. It may also require water overland piping from distant sources.

A 2014 life cycle analysis of natural gas electricity by the National Renewable Energy Laboratory concluded that electricity generated by natural gas from massive hydraulically fractured wells consumed between 249 gallons per megawatt-hour (gal/MWhr) (Marcellus trend) and 272 gal/MWhr (Barnett Shale). The water consumption for the gas from massive hydraulic fractured wells was from 52 to 75 gal/MWhr greater (26 percent to 38 percent greater) than the 197 gal/MWhr consumed for electricity from conventional onshore natural gas.

Some producers have developed hydraulic fracturing techniques that could reduce the need for water. Using carbon dioxide, liquid propane or other gases instead of water have been proposed to reduce water consumption. After it is used, the propane returns to its gaseous state and can be collected and reused. In addition to water savings, gas fracturing reportedly produces less damage to rock formations that can impede production. Recycled flowback water can be reused in hydraulic fracturing. It lowers the total amount of water used and reduces the need to dispose of wastewater after use. The technique is relatively expensive, however, since the water must be treated before each reuse and it can shorten the life of some types of equipment.

Water Contamination

Injected Fluid

In the United States, hydraulic fracturing fluids include proppants, radionuclide tracers, and other chemicals, many of which are toxic. The type of chemicals used in hydraulic fracturing and their properties vary. While most of them are common and generally harmless, some chemicals are carcinogenic. Out of 2,500 products used as hydraulic fracturing additives in the United States, 652 contained one or more of 29 chemical compounds which are either known or possible human carcinogens, regulated under the Safe Drinking Water Act for their risks to human health, or listed as hazardous air pollutants under the Clean Air Act. Another 2011 study identified 632 chemicals used in United States natural gas operations, of which only 353 are well-described in the scientific literature. A study that assessed health effects of chemicals used in fracturing found that 73% of the products had between 6 and 14 different adverse health effects including skin, eye, and sensory organ damage; respiratory distress including asthma; gastrointestinal and liver disease; brain and nervous system harms; cancers; and negative reproductive effects.

An expansive study conducted by the Yale School of Public Health in 2016 found numerous chemicals involved in or released by hydraulic fracturing are carcinogenic. Of the 119 compounds identified in this study with sufficient data, "44% of the water pollutants were either confirmed or possible carcinogens." However, the majority of chemicals lacked sufficient data on carcinogenic potential, highlighting the knowledge gap in this area. Further research is needed to identify both carcinogenic potential of chemicals used in hydraulic fracturing and their cancer risk.

The European Union regulatory regime requires full disclosure of all additives. According to the EU groundwater directive of 2006, "in order to protect the environment as a whole, and human health in particular, detrimental concentrations of harmful pollutants in groundwater must be avoided, prevented or reduced." In the United Kingdom, only chemicals that are "non hazardous in their application" are licensed by the Environment Agency.

Flowback

Less than half of injected water is recovered as flowback or later production brine, and in many cases recovery is <30%. As the fracturing fluid flows back through the well, it consists of spent fluids and may contain dissolved constituents such as minerals and brine waters. In some cases, depending on the geology of the formation, it may contain uranium, radium, radon and thorium. Estimates of the amount of injected fluid returning to the surface range from 15-20% to 30–70%.

Approaches to managing these fluids, commonly known as produced water, include underground injection, municipal and commercial wastewater treatment and discharge, self-contained systems at well sites or fields, and recycling to fracture future wells. The vacuum multi-effect membrane distillation system as a more effective treatment system has been proposed for treatment of flowback. However, the quantity of waste water needing treatment and the improper configuration of sewage plants have become an issue in some regions of the United States. Part of the wastewater from hydraulic fracturing operations is processed there by public sewage treatment plants, which are not equipped to remove radioactive material and are not required to test for it.

Produced water spills and subsequent contamination of groundwater also presents a risk for exposure to carcinogens. Research that modeled the solute transport of BTEX (benzene, toluene, ethylbenzene, and xylene) and naphthalene for a range of spill sizes on contrasting soils overlying groundwater at different depths found that benzene and toluene were expected to reach human health relevant concentration in groundwater because of their high concentrations in produced water, relatively low solid/liquid partition coefficient and low EPA drinking water limits for these contaminants. Benzene is a known carcinogen which affects the central nervous system in the short term and can affect the bone marrow, blood production, immune system, and urogenital systems with long term exposure.

Surface Spills

Surface spills related to the hydraulic fracturing occur mainly because of equipment failure or engineering misjudgments.

Volatile chemicals held in waste water evaporation ponds can evaporate into the atmosphere, or overflow. The runoff can also end up in groundwater systems. Groundwater may become contaminated by trucks carrying hydraulic fracturing chemicals and wastewater if they are involved in accidents on the way to hydraulic fracturing sites or disposal destinations.

In the evolving European Union legislation, it is required that "Member States should ensure that the installation is constructed in a way that prevents possible surface leaks and spills to soil, water or air." Evaporation and open ponds are not permitted. Regulations call for all pollution pathways to be identified and mitigated. The use of chemical proof drilling pads to contain chemical spills is required. In the UK, total gas security is required, and venting of methane is only permitted in an emergency.

Methane

In September 2014, a study from the US 'Proceedings of the National Academy of Sciences' released a report that indicated that methane contamination can be correlated to distance from a well in wells that were known to leak. This however was not caused by the hydraulic fracturing process, but by poor cementation of casings.

Groundwater methane contamination has adverse effect on water quality and in extreme cases may lead to potential explosion. A scientific study conducted by researchers of Duke University found high correlations of gas well drilling activities, including hydraulic fracturing, and methane pollution of the drinking water. According to the 2011 study of the MIT Energy Initiative, "there is evidence of natural gas (methane) migration into freshwater zones in some areas, most likely as a result of substandard well completion practices i.e. poor quality cementing job or bad casing, by a few operators." A 2013 Duke study suggested that either faulty construction (defective cement seals in the upper part of wells, and faulty steel linings within deeper layers) combined with a peculiarity of local geology may be allowing methane to seep into waters; the latter cause may also release injected fluids to the aquifer. Abandoned gas and oil wells also provide conduits to the surface in areas like Pennsylvania, where these are common.

A study by Cabot Oil and Gas examined the Duke study using a larger sample size, found that methane concentrations were related to topography, with the highest readings found in low-lying areas, rather than related to distance from gas production areas. Using a more precise isotopic analysis, they showed that the methane found in the water wells came from both the formations where hydraulic fracturing occurred, and from the shallower formations. The Colorado Oil & Gas Conservation Commission investigates complaints from water well owners, and has found some wells to contain biogenic methane unrelated to oil and gas wells, but others that have thermogenic methane due to oil and gas wells with leaking well casing. A review published in February 2012 found no direct evidence that hydraulic fracturing actual injection phase resulted in contamination of ground water, and suggests that reported problems occur due to leaks in its fluid or waste storage apparatus; the review says that methane in water wells in some areas probably comes from natural resources.

Another 2013 review found that hydraulic fracturing technologies are not free from risk of contaminating groundwater, and described the controversy over whether the methane that has been detected in private groundwater wells near hydraulic fracturing sites has been caused by drilling or by natural processes.

Radionuclides

There are naturally occurring radioactive materials (NORM), for example radium, radon, uranium, and thorium, in shale deposits. Brine co-produced and brought to the surface along with the oil and gas sometimes contains naturally occurring radioactive materials; brine from many shale gas wells, contains these radioactive materials. The U.S. Environmental Protection Agency and regulators in North Dakota consider radioactive material in flowback a potential hazard to workers at hydraulic fracturing drilling and waste disposal sites and those living or working nearby if the correct procedures are not followed. A report from the Pennsylvania Department of Environmental Protection indicated that there is little potential for radiation exposure from oil and gas operations.

Land Usage

In the UK, the likely well spacing visualised by the December 2013 DECC Strategic Environmental Assessment report indicated that well pad spacings of 5 km were likely in crowded areas, with up to 3 hectares (7.4 acres) per well pad. Each pad could have 24 separate wells. This amounts to 0.16% of land area. A study published in 2015 on the Fayetteville Shale found that a mature gas field impacted about 2% of the land area and substantially increased edge habitat creation. Average land impact per well was 3 hectares (about 7 acres) Research indicates that effects on ecosystem services costs (i.e. those processes that the natural world provides to humanity)has reached over $250 million per year in the U.S.

Seismicity

Hydraulic fracturing causes induced seismicity called microseismic events or micro-earthquakes. These microseismic events are often used to map the horizontal and vertical extent of the fracturing. The magnitude of these events is usually too small to be detected at the surface, although the biggest micro-earthquakes may have the magnitude of about -1.5 (M_w).

Induced Seismicity from Hydraulic Fracturing

As of August 2016, there were at least nine known cases of fault reactivation by hydraulic fracturing that caused induced seismicity strong enough to be felt by humans at the surface: In Canada, there have been three in Alberta (M 4.8 and M 4.4 and M 4.4) and three in British Columbia (M 4.6, M 4.4 and M 3.8); In the United States there has been: one in Oklahoma (M 2.8) and one in Ohio (M 3.0), and; In the United Kingdom, there have been two in Lancashire (M 2.3 and M 1.5).

Induced Seismicity from Water Disposal Wells

According to the USGS only a small fraction of roughly 30,000 waste fluid disposal wells for oil and gas operations in the United States have induced earthquakes that are large enough to be of concern to the public. Although the magnitudes of these quakes has been small, the USGS says that there is no guarantee that larger quakes will not occur. In addition, the frequency of the quakes has been increasing. In 2009, there were 50 earthquakes greater than magnitude 3.0 in the area spanning Alabama and Montana, and there were 87 quakes in 2010. In 2011 there were 134 earthquakes in the same area, a sixfold increase over 20th century levels. There are also concerns that quakes may damage underground gas, oil, and water lines and wells that were not designed to withstand earthquakes.

A 2012 US Geological Survey study reported that a "remarkable" increase in the rate of M ≥ 3 earthquakes in the US midcontinent "is currently in progress", having started in 2001 and culminating in a 6-fold increase over 20th century levels in 2011. The overall increase was tied to earthquake increases in a few specific areas: the Raton Basin of southern Colorado (site of coalbed methane activity), and gas-producing areas in central and southern Oklahoma, and central Arkansas. While analysis suggested that the increase is "almost certainly man-made", the USGS noted: "USGS's studies suggest that the actual hydraulic fracturing process is only very rarely the direct cause of felt earthquakes." The increased earthquakes were said to be most likely caused by increased injection of gas-well wastewater into disposal wells. The injection of waste water from oil and gas operations, including from hydraulic fracturing, into saltwater disposal wells may cause bigger low-magnitude tremors, being registered up to 3.3 (M_w).

Noise

Each well pad (in average 10 wells per pad) needs during preparatory and hydraulic fracturing process about 800 to 2,500 days of activity, which may affect residents. In addition, noise is created by transport related to the hydraulic fracturing activities. Noise pollution from hydraulic fracturing operations (e.g., traffic, flares/burn-offs) is often cited as a source of psychological distress, as well as poor academic performance in children. For example, the low-frequency noise that comes from well pumps contributes to irritation, unease, and fatigue.

The UK Onshore Oil and Gas (UKOOG) is the industry representative body, and it has published a charter that shows how noise concerns will be mitigated, using sound insulation, and heavily silenced rigs where this is needed.

Community Impacts

Impacted communities are often already vulnerable, including poor, rural, or indigenous persons, who may continue to experience the deleterious effects of hydraulic fracturing for generations. Competition for resources between farmers and oil companies contributes to stress for agricultural workers and their families, as well as to a community-level "us versus them" mentality that creates community distress. Rural communities that host hydraulic fracturing operations often experience a "boom/bust cycle," whereby their population surges, consequently exerting stress on community infrastructure and service provision capabilities (e.g., medical care, law enforcement).

Indigenous and agricultural communities may be particularly impacted by hydraulic fracturing, given their historical attachment to, and dependency on, the land they live on, which is often damaged as a result of the hydraulic fracturing process. Native Americans, particularly those living on rural reservations, may be particularly vulnerable to the effects of fracturing; that is, on the one hand, tribes may be tempted to engage with the oil companies to secure a source of income but, on the other hand, must often engage in legal battles to protect their sovereign rights and the natural resources of their land.

Environmental Impact of the Oil Shale Industry

Environmental impact of the oil shale industry includes the consideration of issues such as land use, waste management, and water and air pollution caused by the extraction and processing of oil shale. Surface mining of oil shale deposits causes the usual environmental impacts of open-pit mining. In addition, the combustion and thermal processing generate waste material, which must be disposed of, and harmful atmospheric emissions, including carbon dioxide, a major greenhouse gas. Experimental in-situ conversion processes and carbon capture and storage technologies may reduce some of these concerns in future, but may raise others, such as the pollution of groundwater.

Kiviõli Oil Shale Processing & Chemicals Plant in Ida-Virumaa, Estonia.

Surface Mining and Retorting

Land use and Waste Management

Surface mining and *in-situ* processing requires extensive land use. Mining, processing and waste disposal require land to be withdrawn from traditional uses, and therefore should avoid high density population areas. Oil shale mining reduces the original ecosystem diversity with habitats supporting a variety of plants and animals. After mining the land has to be reclaimed. However, this process takes time and cannot necessarily re-establish the original biodiversity. The impact of sub-surface mining on the surroundings will be less than for open pit mines. However, sub-surface mining may also cause subsidence of the surface due to the collapse of mined-out area and abandoned stone drifts.

Disposal of mining wastes, spent oil shale (including semi-coke) and combustion ashes needs additional land use. According to the study of the European Academies Science Advisory Council, after processing, the waste material occupies a greater volume than the material extracted, and therefore cannot be wholly disposed underground. According to this, production of a barrel of shale oil can generate up to 1.5 tonnes of semi-coke, which may occupy up to 25% greater volume than the original shale. This is not confirmed by the results of Estonia's oil shale industry. The mining and processing of about one billion tonnes of oil shale in Estonia has created about 360-370 million tonnes of solid waste, of which 90 million tonnes is a mining waste, 70–80 million tonnes is a semi-coke, and 200 million tonnes are combustion ashes.

The waste material may consist of several pollutants including sulfates, heavy metals, and polycylic aromatic hydrocarbons (PAHs), some of which are toxic and carcinogenic. To avoid contamination of the groundwater, the solid waste from the thermal treatment process is disposed in an open dump (landfill or "heaps"), not underground. As semi-coke consists of, in addition to minerals, up to 10% organics that may pose hazard to the environment owing to leaching of toxic compounds as well as to the possibility of self-ignition.

Water Management

Mining influences the water runoff pattern of the area affected. In some cases it requires the lowering of groundwater levels below the level of the oil shale strata, which may have harmful effects on the surrounding arable land and forest. In Estonia, for each cubic meter of oil shale mined, 25 cubic meters of water must be pumped from the mine area. At the same time, the thermal processing of oil shale needs water for quenching hot products and the control of dust. Water concerns are particularly sensitive issue in arid regions, such as the western part of the United States and Israel's Negev Desert, where there are plans to expand the oil shale industry. Depending on technology, above-ground retorting uses between one and five barrels of water per barrel of produced shale oil. *In situ* processing, according to one estimate, uses about one-tenth as much water.

Water represents the major vector of transfer of oil shale industry pollutants. One environmental issue is to prevent noxious materials leaching from spent shale into the water supply. The oil shale processing is accompanied by the formation of process waters and waste waters containing phenols, tar and several other products, heavily separable and toxic to the environment. A 2008 programmatic environmental impact statement issued by the United States Bureau of Land Management stated that surface mining and retort operations produce 2 to 10 U.S. gallons (7.6 to 37.9 l; 1.7 to 8.3 imp gal) of waste water per 1 short ton (0.91 t) of processed oil shale.

Air Pollution Management

Main air pollution is caused by the oil shale-fired power plants, which provide the atmospheric emissions of gaseous products like nitrogen oxides, sulfur dioxide and hydrogen chloride, and the airborne particulate matter (fly ash). It includes particles of different types (carbonaceous, inorganic ones) and different sizes. The concentration of air pollutants in flue gas depends primarily on the combustion technology and burning regime, while the emissions of solid particles are determined by the efficiency of fly ash-capturing devices.

Open deposition of semi-coke causes distribution of pollutants in addition to aqueous vectors also via air (dust).

There are possible links from being in an oil shale area to a higher risk of asthma and lung cancer than other areas.

Greenhouse Gas Emissions

Carbon dioxide emissions from the production of shale oil and shale gas are higher than conventional oil production and a report for the European Union warns that increasing public concern about the adverse consequences of global warming may lead to opposition to oil shale development.

Emissions arise from several sources. These include CO_2 released by the decomposition of the kerogen and carbonate minerals in the extraction process, the generation of the energy needed to heat the shale and in the other oil and gas processing operations, and fuel used in the mining of the rock and the disposal of waste. As the varying mineral composition and calorific value of oil shale deposits varies widely, the actual values vary considerably. At best, the direct combustion of oil shales produces carbon emissions similar to those from the lowest form of coal, lignite, at 2.15 moles CO_2/MJ, an energy source which is also politically contentious due to its high emission levels. For both power generation and oil extraction, the CO_2 emissions can be reduced by better utilization of waste heat from the product streams.

In-situ Processing

Currently, the *in-situ* process is the most attractive proposition due to the reduction in standard surface environmental problems. However, *in-situ* processes do involve possible significant environmental costs to aquifers, especially since *in-situ* methods may require ice-capping or some other form of barrier to restrict the flow of the newly gained oil into the groundwater aquifers. However, after the removal of the freeze wall these methods can still cause groundwater contamination as the hydraulic conductivity of the remaining shale increases allowing groundwater to flow through and leach salts from the newly toxic aquifer.

Environmental Impact of the Petroleum Industry

Impacts of Drilling for Oil

Petroleum companies drill oil from the ground by use of drilling rigs and wells that reaches the pocket of oil bed. The oil is likely to fill the rock layer and spread throughout the open place. The oil may spill into the water bodies like oceans, lakes, or rivers. The crude oil contains toxic material that is lethal to water animals like fish and reptiles and may lead to several deaths within a short time. Distillates of the crude oil and petroleum may also cause congenital disabilities to fish and other animals which come into contact with them. Drilling for oil also requires that the ecology of an area be altered by clearing land to create adequate space for drilling of oil.

Petroleum refinery is a major source of pollution in areas where they are established. The refineries are major sources of toxic air pollutants including BTEX compounds, carbon monoxide, particulate matters, and sulfur dioxide. Some of toxic chemicals released into the air are suspect cancer-causing agents and are also responsible for the development of reproductive problems, and respiratory complications. A large amount of carbon monoxide emissions traps the heat in the earth leading to climate change.

Petroleum refineries are also major contaminators of surface and ground water. The deep wells for the disposal of waste material end up in aquifers and ground water. Some of the refineries also discharge untreated waste material into the water bodies such as lakes and rivers. This means of waste disposal into the rivers affects the quality of water and the water animals. The petroleum products that find their ways into the water bodies are also highly inflammable and may cause river fires like has been the case of Cuyahoga River. Petroleum refining activities may also contaminate the soil. Soil contamination includes the hazardous waste, oil spills, sludge from the treatment process, and coke dust. Soil contamination reduces the fertility of the soil and introduces foreign particles which may affect the growth and quality of crops.

References

- Hass, Benjamin (14 August 2012). "Fracking Hazards Obscured in Failure to Disclose Wells". Bloomberg News. Retrieved 27 March 2013

- Taherdangkoo, Reza; Tatomir, Alexandru; Anighoro, Tega; Sauter, Martin (February 2019). "Modeling fate and transport of hydraulic fracturing fluid in the presence of abandoned wells". Journal of Contaminant Hydrology. 221: 58–68. Doi:10.1016/j.jconhyd.2018.12.003

- What-is-the-environmental-impact-of-the-petroleum-industry: worldatlas.com, Retrieved 7 March, 2019

- Kattel, T. (2003). "Design of a new oil shale surface mine"(PDF). Oil Shale. A Scientific-Technical Journal. Estonian Academy Publishers. 20 (4): 511–514. ISSN 0208-189X. Retrieved 23 June 2007

Permissions

All chapters in this book are published with permission under the Creative Commons Attribution Share Alike License or equivalent. Every chapter published in this book has been scrutinized by our experts. Their significance has been extensively debated. The topics covered herein carry significant information for a comprehensive understanding. They may even be implemented as practical applications or may be referred to as a beginning point for further studies.

We would like to thank the editorial team for lending their expertise to make the book truly unique. They have played a crucial role in the development of this book. Without their invaluable contributions this book wouldn't have been possible. They have made vital efforts to compile up to date information on the varied aspects of this subject to make this book a valuable addition to the collection of many professionals and students.

This book was conceptualized with the vision of imparting up-to-date and integrated information in this field. To ensure the same, a matchless editorial board was set up. Every individual on the board went through rigorous rounds of assessment to prove their worth. After which they invested a large part of their time researching and compiling the most relevant data for our readers.

The editorial board has been involved in producing this book since its inception. They have spent rigorous hours researching and exploring the diverse topics which have resulted in the successful publishing of this book. They have passed on their knowledge of decades through this book. To expedite this challenging task, the publisher supported the team at every step. A small team of assistant editors was also appointed to further simplify the editing procedure and attain best results for the readers.

Apart from the editorial board, the designing team has also invested a significant amount of their time in understanding the subject and creating the most relevant covers. They scrutinized every image to scout for the most suitable representation of the subject and create an appropriate cover for the book.

The publishing team has been an ardent support to the editorial, designing and production team. Their endless efforts to recruit the best for this project, has resulted in the accomplishment of this book. They are a veteran in the field of academics and their pool of knowledge is as vast as their experience in printing. Their expertise and guidance has proved useful at every step. Their uncompromising quality standards have made this book an exceptional effort. Their encouragement from time to time has been an inspiration for everyone.

The publisher and the editorial board hope that this book will prove to be a valuable piece of knowledge for students, practitioners and scholars across the globe.

Index

A

Aromatic Hydrocarbons, 23, 198, 204, 216, 220, 223, 231

Artificial Lift, 27, 55-57

Asphalt, 2, 4, 38, 132, 136-137, 164, 189

B

Bakken Formation, 71, 73

Beam Pumping, 27, 116

Bitumen, 1-2, 5, 7, 12, 38-39, 98, 101-102, 154, 183, 187

C

Cap Rock, 9-10, 20, 130

Cationic Membrane, 83, 87-88

Clay Minerals, 7-8, 83, 96, 104

Conventional Drilling, 39, 80

Crude Oil Distillation Unit, 138, 141, 160-162

D

Deepwater Drilling, 50, 104-106, 120

Directional Drilling, 47-48, 107-111

Drilling Platform, 47-49, 52

Drilling Rig, 44, 50, 64-65, 78, 104, 108, 132, 165

E

Electrical Submersible Pumps, 27, 57

Electrochemical Potential, 84, 97

Electrokinetic Potential, 84

Ethylene Glycol, 75-76, 171, 217

Exploration Well, 31, 33, 36, 65, 125

F

Fluid Catalytic Cracking, 134, 161, 168

Formation Pressure, 55, 93, 157, 217

Fractional Distillation, 23, 31, 101, 135, 189, 198

G

Gamma Ray Logging, 53, 102-103

Gas Cycling, 60

Gaseous Hydrocarbons, 1, 38, 158

Geological Mapping, 34

Gravitational Force, 41, 81

H

Hubbert Peak Theory, 1, 21

Hydraulic Fracturing, 13, 22, 27, 31, 54-55, 70-77, 79-81, 165, 213, 218, 222-230, 234

Hydrocarbon Content, 3

Hydrocarbon Exploration, 65, 67

Hydrocarbon Saturation, 68, 93

Hydrogen Index, 89, 91

I

Isooctane, 22, 139, 199-200

K

Kerogen, 6-8, 12-13, 150, 153, 232

L

Liquefied Petroleum Gas, 60, 62, 74, 132, 159, 167, 175, 177, 202

Liquid Petroleum, 2, 38, 175, 188

Liquid-junction Potential, 84

Logging Tool, 36, 52, 89, 93, 96

Logging-while-drilling, 48, 53, 119

M

Measurement-while-drilling, 53, 95, 119

Methane, 4, 7, 14, 22, 71, 80, 107, 169, 181, 216, 221-224, 227-229

Microseismic Monitoring, 77

Mud-filtrate Salinity, 82, 85

N

Natural Gas, 1-2, 7-9, 14, 19, 22, 27, 30, 38, 40, 58-62, 65, 73, 77, 105, 114, 116, 127, 129-130, 139, 155, 161, 165, 176, 178, 187, 189, 191, 221, 227

Natural Gas Liquids, 7, 60-61, 152, 168

Neutron Porosity, 31, 53, 89-91, 96

O

Octane Rating, 1, 22, 160, 179, 200, 204, 206-207

Offshore Drilling, 22, 49, 51, 105
Oil Refinery, 129, 132-134, 136, 140, 143, 163, 186, 198-199

P
Paraffin, 4-5, 7, 101, 136, 167, 193, 196, 199-200, 211-212
Petroleum Exploration, 15, 31, 33-34, 39, 42
Petroleum Geology, 1, 20-21, 65
Petroleum Production, 13-14, 25-27, 38-39, 50, 52, 62, 167, 213
Petroleum Reservoir, 47, 155
Pore Fluid Pressure, 72
Pore Pressure Gradient, 111-112
Porosity, 9, 11, 13, 18, 31, 53-54, 67, 84-85, 89-91, 94-98, 104, 119, 156-157
Pressure Safety Valves, 145, 147, 149
Production Tubing, 53-54, 56, 64
Pumpjack, 113-114, 116, 199

R
Rate of Penetration, 48, 100, 111, 125
Reservoir Engineering, 25-26, 29, 39, 124
Resistivity Log, 94-95
Rotary Steerable System, 45
Rotary Steerable Systems, 53, 109

S
Sand Control, 27-29, 111
Seal Rock, 18, 21
Seismic Acquisition, 42-43
Seismic Activity, 71-72, 218
Shale Gas, 14, 54, 68, 70, 73, 80-81, 153-154, 165, 224, 228, 232
Sodium Chloride, 4, 75, 83, 141, 162
Spectral Noise Logging, 98
Stratigraphic Traps, 10-11, 18, 21, 41
Strip Mining, 151
Sucker Rod, 113, 115

T
Tension-leg Platform, 50, 106
Thermal Cracking, 134

V
Vacuum Distillation, 102, 138-140, 159-161, 193

W
Well Logging, 36, 52-53, 92-93, 95, 99, 102, 124, 128, 132

X
Xylene, 23, 168, 175, 200, 226